SUSTAINABLE AGRICULTURE AND RESISTANCE

SUSTAINABLE AGRICULTURE AND RESISTANCE

Transforming Food Production in Cuba

EDITED BY

Fernando Funes, Luis García,
Martin Bourque, Nilda Pérez, and Peter Rosset

Co-published with ACTAF (Asociación Cubana de Técnicos Agrícolas y Forestales) and CEAS (Centro de Estudios de Agricultura Sostenible, Universidad Agraria de La Habana)

FOOD FIRST BOOKS
Oakland, California

Translated from the Spanish: *Transformando el Campo Cubano: Avances de la Agricultura Sostenible,* copyright © 2001 Havana, Cuba, Asociación Cubana de Técnicos Agrícolas y Forestales (ACTAF).

Co-published with Asociación Cubana de Técnicos Agrícolas y Forestales (ACTAF) and Centro de Estudios de Agricultura Sostenible CEAS.

Text design by Jeff Brandenburg/ImageComp
Cover design by Colored Horse Studios
Cover photograph by Peter Rosset of a mural in the lobby of the Agrarian University of Havana (UNAH)
Map by Michael Manoochehri
Translation by Dulce María Vento Cárdenas, Lidia González Seco, Robin Clement, Kristen Cañizares, and Peter Rosset

Library of Congress Cataloging-in-Publication Data

Transformando el campo Cubano. English.
Sustainable agriculture and resistance : transforming food production in Cuba /edited by Fernando Funes . . . [et al. ; translation by Dulce María . . . et al.].
p. cm.
Includes bibliographical references (p.).
ISBN: 0-935028-87-0
1. Sustainable agriculture—Cuba. 2. Organic farming—Cuba. 3. Sustainable development—Cuba. I. Funes, Fernando. II. Title.
S477.C8 T7313 2002
338.1'62'097291—dc21 2001058609

Food First Books are distributed by:
CDS
425 Madison Avenue
New York, NY 10017
(800) 343-4499

PRINTED IN CANADA
10 9 8 7 6 5 4 3 2

Dedication

Dedicated to the memory of our compañeros Manuel Álvarez Pinto, Jorge Ramon Cuevas, and Yanet Ojeda Hernández, representing three generations of struggle for agroecology and the environment.

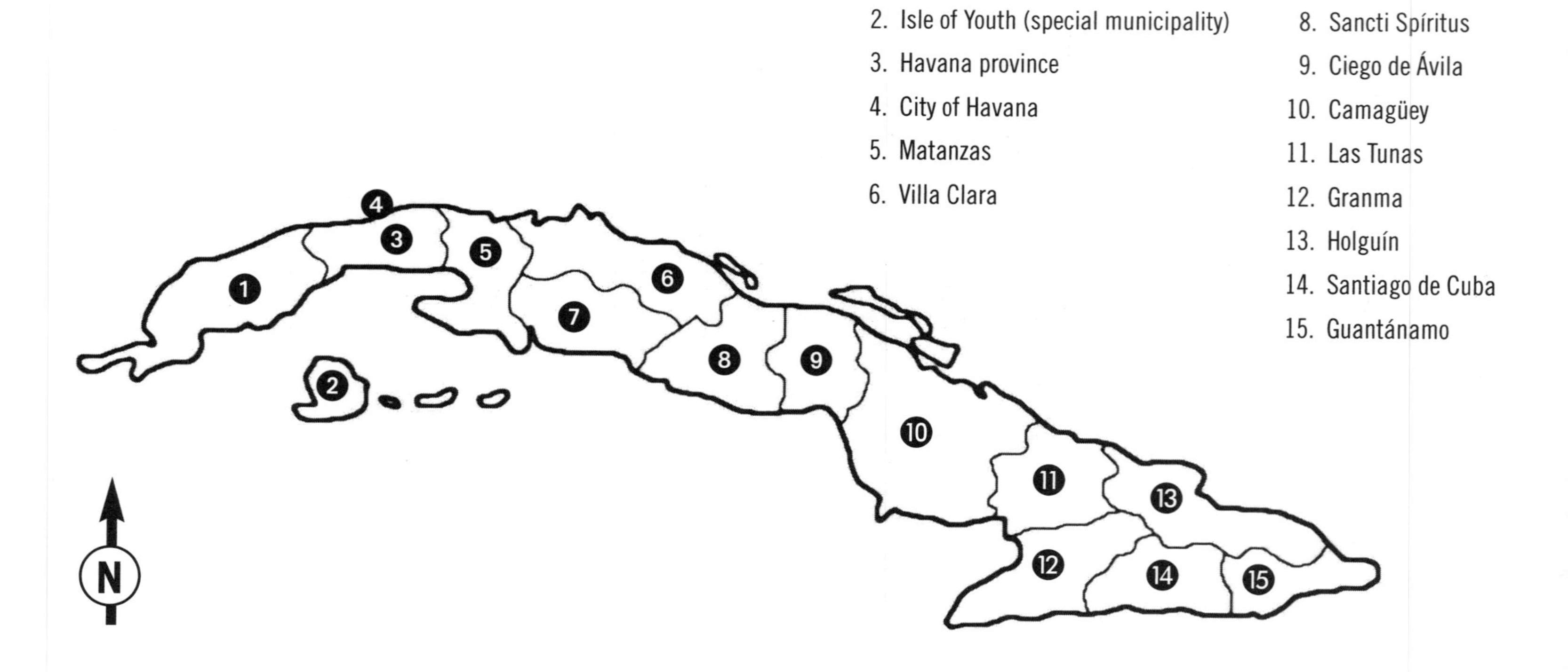
THE PROVINCES OF CUBA
1. Pinar del Río
2. Isle of Youth (special municipality)
3. Havana province
4. City of Havana
5. Matanzas
6. Villa Clara
7. Cienfuegos
8. Sancti Spíritus
9. Ciego de Ávila
10. Camagüey
11. Las Tunas
12. Granma
13. Holguín
14. Santiago de Cuba
15. Guantánamo
N

TABLE OF CONTENTS

Note from the Editors

"Make human life more rational. Build a just international economic order. Use all science for a more sustainable development that does not contaminate the environment. Pay the ecological debt and not the external debt. Fight hunger, not people."

—Fidel Castro[1]

For more than a decade, Cuba—a small, underdeveloped, Third World country—suffered a terrible economic crisis. This crisis was brought on by the tightening of the unjust US trade embargo, which has lasted more than 40 years, and by the disintegration of the USSR and the collapse of the Socialist Bloc.

To resist this crisis, Cuba has been slowly transforming her intensive, industrial, "Green Revolution," methods of food and fiber production into a more sustainable kind of agriculture, characterized by self-reliance, low input use, and organic farming practices. The building blocks of this new model of agriculture have been various: traditional farming knowledge, preserved for generations by Cuban farmers; the accumulated research results of Cuban laboratories and experiment stations; the highly educated populace of this island nation; and the priority that the Cuban government has placed, for years, on developing alternative practices. The impact of these changes is just beginning to be felt in the recovery and advancement of Cuban agriculture.

The earliest advances with organic farming practices in Cuba were first publicized to the world in 1994 in *The Greening of the Revolution*,[2] coauthored by one of the present editors. The book sparked international interest by showing what a small country could do to guarantee the health and food security of its population while protecting its environment, under the most difficult of circumstances. It also showed what could be done with little more than dedication, intelligence, and minimal resources. Much progress has been made in subsequent years, and thus we bring you this new volume, written in great part by Cuban authors.

Part I of this book provides an historical and contemporary overview of the social, economic, nutritional, philosophical, institutional, and educational context of Cuban agriculture and its recent transformation. Part II covers the agronomic practices that characterize the new model of food production in Cuba, ranging from pest management and intercropping to animal traction, soil husbandry, and the integration of crops and livestock. Part III covers specific products, crops, farming systems and sectors, from "green" medicine and urban agriculture to tobacco and sugarcane.

In the discussion of alternative practices such as ecological pest management, intercropping, and the integration of plants and animals, one can see the emergence of a new holistic, agroecological paradigm. This new model offers important lessons about biological efficiency, ecological economy, energetics, and the environment, which merit reflection and analysis for the future.

We give deep thanks to our collaborators in many Cuban institutions, for the patient work of writing, editing, and refereeing the chapters of this book. We do not list them here for fear of omissions. This book is part of a collaborative project of the Institute for Food and Development Policy/Food First, supported by the C.S. Mott Foundation, the Pond Foundation, the Boehm Foundation, the John D. and Catherine T. MacArthur Foundation, the Ford Foundation and by thousands of individual members of Food First, whom we thank for making this possible.

Finally, we want to acknowledge that this wealth of information about sustainable agriculture, and resistance in the face of adversity, would not exist were it not for the abnegation and daily sacrifice of thousands and thousands of Cuban farmers, technicians, scientists, and professors, and the willingness to change shown by the Cuban government under such difficult circumstances.

—The Editors

Notes

1. Fidel Castro, speech given at the United Nations Conference on Environment and Development, Rio de Janeiro, Brazil, 1992.

2. Rosset, P.M., and M. Benjamin (eds.). 1994. *The Greening of the Revolution: Cuba's Experiment with Organic Agriculture.* Australia: Ocean Press.

PROLOGUE

The Principles and Strategies of Agroecology in Cuba

Miguel A. Altieri, University of California at Berkeley

The concept of sustainable agriculture comes as a relatively recent response to the decline in the quality of the natural resource base linked to modern agriculture. This decline in the potential for future productivity has prompted a great deal of discussion about the need to make major adjustments to conventional agriculture, to make it more environmentally sound, socially just, and economically viable. Research on alternatives to high-input systems has focused on the reduction, or total elimination, of agrochemicals. This can be accomplished through changes in system design and management to ensure proper plant nutrition and protection, using sources of organic matter, and alternative pest management techniques. In developing countries, an important concern is to ensure that alternative technologies offer real possibilities for small and resource-poor farmers, while enhancing household food security, and protecting the environment.

The Cuban experience has demonstrated that adoption of agroecological methods in rural communities (and in cities, through urban agriculture), can bring about productive and economic benefits in a socially equitable manner. At the beginning of the economic crisis brought on by the collapse of trade with the former Socialist Bloc, the alternatives implemented in Cuban agriculture emphasized an input substitution approach, relying heavily on locally produced biopesticides and biofertilizers to make up for the lack of imported inputs. The prevailing philosophy saw pests and nutrient deficiencies as the main causes of low productivity. Overcoming such limiting factors by adding external organic inputs was seen as the way to stem yield losses. This persistent viewpoint at first prevented Cuban agronomists from understanding that limiting factors are but mere symptoms of a much deeper ecological disease of agroecosystems, caused by the imbalances inherent in large-scale monoculture and high-input regimes.

Thanks to the training and research efforts of government and non-government organizations, this reductionist approach has gradually started to change on the island of Cuba. This is in part thanks to the pioneering work of organizations like the Cuban Organic Agriculture Association (ACAO)—now called the Organic Farming Group (GAO) of the Cuban Association of Agriculture and Forestry Technicians (ACTAF)—, the National Association of Small Farmers (ANAP), the Fundación Antonio Núñez Jiménez de la Naturaleza y el Hombre, and the Department of Coordination and Project Assistance of the Council of Churches of Cuba (DECAP-CIC). This evolution has provided agroecological knowledge to thousands of Cuban farmers and technicians, thereby giving them the tools to apply ecological principles to the design and management of diverse and sustainable agroecosystems. Agroecology goes well beyond the use of low-input technologies, to minimize dependency on external inputs. The emphasis is on the design of complex agroecosystems that take advantage of ecological interactions, and synergisms between biotic and abiotic components—mechanisms by which soil fertility enhancement, biological pest control, and higher productivity can be achieved through internal processes.

In Cuba today, hundreds of farms are managed using the concepts of agroecology. The success of such farms is the direct result of the application of several ecological principles. These include:

- increased recycling of nutrients and biomass within the system
- addition of organic matter to improve soil quality and activate soil biology
- soil and water conservation to minimize resource losses
- diversification of agroecosystems in time and space, including the integration of crops and livestock
- optimization of biological interactions and synergisms among functional components of biodiversity, to provide key ecological services
- integration of farm components to increase biological efficiencies and preserve the productive capacity of the agroecosystem

The diversification of agroecosystems is a key strategy, and to achieve this Cubans have used a wide variety of techniques such as crop rotations, green manures, polycultures, agroforestry, and crop–livestock integration. Agroecologists now realize that these diversified systems mimic natural ecological processes, and that the sustainability of such systems is based on the very ecological models that they follow. Like their natural analogs, diversified cropping systems optimize nutrient cycling and organic matter accumulation, close energy flows, conserve resources, exhibit immunity to pests, and sustain levels of productivity over time.

The challenge is to discover the most efficient crop, tree, and animal combinations that match the environmental potential of each area. The advances achieved by agroecology in Cuba to date, and which are reflected in this book, are a valuable set of lessons for the millions of people throughout the world who are committed to a truly sustainable agriculture.

As one passes through the diverse chapters, one can glean numerous critically important agroecological concepts and principles. Among these are:

- optimization of local resources and promotion of *within-farm* synergisms through plant–animal combinations
- reliance on the ecological services of biodiversity in order to minimize the use of external inputs, whether organic or conventional
- matching of cropping systems with existing soil and climatic potentials
- conservation and use of crop and non-crop biodiversity within and around farms to maximize utilization of biological and genetic resources
- reliance on the knowledge and wisdom of locals and farmers as a key input
- promotion of participatory methods in research and in the extension and implementation process

Without a doubt the Cuban experience demonstrates that agroecology is a powerful science which can provide guidelines on how to develop systems that take full advantage of biodiversity and natural processes. The correct spatial and temporal integration of plants and animals maximizes interactions and synergisms. Often this will lead to improved system function. The final design, however, must be adapted to local socioeconomic circumstances and biophysical constraints. The resultant systems must meet the needs and aspirations of the rural population.

Within the political and economic context of the difficult situation facing Cuba today, agroecology has served as an important means of resistance—a source of knowledge to promote farming under conditions of externally-imposed, severe input shortages. The impact of this experience surely will dramatically influence the way Cuba manages its agricultural resources well into the twenty-first century, long after today's circumstances have passed into history.

INTRODUCTION

Lessons of Cuban Resistance

Peter Rosset, Institute for Food and Development Policy/Food First
and Martin Bourque, the Ecology Center

At the Institute for Food and Development Policy (Food First), we have spent 25 years studying hunger around the world, and its relationship to agriculture and rural development. Over the years we have seen many countries enter into food crises. The proximate causes have been many, ranging from wars to droughts or floods, though invariably the ultimate causes have been more tightly linked to inequality or lack of social justice in some form, be it in access to land, jobs, government assistance, or the structure of the world economy. Such crises have often led to famines, which have only been resolved by massive international intervention and food aid, usually leaving the afflicted region or country less able to feed itself in the future and more dependent on food imports from the West than ever before (Lappé and Collins 1977; Lappé et al. 1998).

The experience of Cuban resistance to external shocks during the 1990s stands in sharp contrast to this panorama. When collapsing trade relations plunged this island nation into a food crisis, foreign assistance and food aid were scarcely available, thanks to the tightening of the US trade embargo. Cuba was forced to turn inward, toward its own natural and human resources, and tap both old and new ways to boost production of basic foods without relying on imports. It wasn't easy, but in many ways the Cuban people and government were uniquely prepared to resist—the well-educated and energetic populace put their dynamism and ingenuity to the task, and the government its commitment to food for all and its support for domestic science and technology. Cubans and their government pulled through the crisis, and their tale offers powerful lessons about resistance, self-reliance, and alternative policies and production methods that could well serve other countries facing their own rural and food crises (Rosset 1998; Rosset and Altieri 1994).

Our global food system is in the midst of a multifaceted crisis, with ecological, economic, and social dimensions. To overcome that crisis, political and social changes are needed to allow the widespread development of alternatives. While there are many examples of farmer-driven and community-based alternative agricultural development models throughout the world that demonstrate that the alternatives work and are economically viable, Cuba offers one of the few examples where fundamental policy shifts and serious governmental resources have supported this movement. It is therefore important all people interested in developing food systems that are socially just, environmentally sustainable, and economically viable, pay close attention to current policy and technological developments in Cuba.

The world food system, based largely on the conventional "Green Revolution" model, is productive—there should be no doubt about that—as per capita food produced in the world has increased by 15 percent over the past thirty-five years. But that production is in ever fewer hands, and costs ever more in economic and ecological terms. In spite of these per capita increases, and an overall per capita surplus of calories, protein, and fats, there are at least 800 million people in the world who do not reap adequate benefits from this production. And it is getting worse. In the last twenty years the number of hungry people in the world—excluding China—has risen by 60 million (Lappé et al. 1998).

Ecological impacts of industrial-style farming include adverse effects on groundwater due to abuse of irrigation and to pesticide and fertilizer runoff, on biodiversity through the spread of large scale monocultures and the elimination of traditional crop varieties, and on the very capacity of agroecosystems to be productive into the future (Rosset and Altieri 1997; Rosset 1997a).

Economically, production costs rise as farmers are forced to use ever more expensive machines and farm chemicals, while crop prices continue a several-decade-long downward trend, causing a cost-price squeeze which led to the loss of untold tens of millions of farmers worldwide through bankruptcies. Socially, we have the concentration of farmland in fewer and fewer hands as low crop prices make farming on a small scale unprofitable (despite higher per acre total productivity of small farms), and agribusiness corporations extend their control over more and more basic commodities (Rosset and Altieri 1997).

Clearly the dominant corporate food system is not capable of adequately addressing the needs of people or of the environment. Yet there are substantial obstacles to the widespread adoption of alternatives. The greatest obstacles are presented by political-corporate power and vested interests, yet at times the psychological barrier to believing that the alternatives can work seems almost as difficult to overcome. The oft-repeated challenge is: "Could organic farming (or agroecology, local production, small farms, farming without pesticides) ever really feed the entire population of a country?" (Rosset 1999). Recent Cuban history—the overcoming of a food crisis through self-reliance, smaller farms, and agroecological technology—shows us that the alternatives can indeed feed a nation, and thus provides a crucial case study for the ongoing debate.

Around the globe, farmers, activists, and researchers are working to create a new model for agriculture that responds to the multiple facets of the crisis. The goals of this model are to be environmentally sound, economically viable, socially just, and culturally appropriate. The Cuban experience presented in this volume offers many new ideas to this movement.

A Brief History

When trade relations with the Soviet Bloc crumbled in late 1989 and 1990, and the US tightened the trade embargo, Cuba was plunged into economic crisis. In 1991 the government declared the "Special Period in Peacetime," which basically put the country on a wartime economy-style austerity program. An immediate 53 percent reduction in oil imports not only affected fuel availability for the economy, but also reduced to zero the foreign exchange that Cuba had formerly obtained via the re-export of petroleum. Imports of wheat and other grains for human consumption dropped by more than 50 percent, while other foodstuffs declined even more. Cuban agriculture was faced with an initial drop of about 70 percent in the availability of fertilizers and pesticides, and more than 50 percent in fuel and other energy sources produced by petroleum (Rosset and Benjamin 1994).

Suddenly, a country with an agricultural sector technologically similar to California's found itself almost without chemical inputs, with sharply reduced access to fuel and irrigation, and with a collapse in food imports. In the early 1990s average daily caloric and protein intake by the Cuban population may have been as much as 30 percent below levels in the 1980s.

Fortunately, Cuba was not totally unprepared to face the critical situation that arose after 1989. It had, over the years, emphasized the development of human resources, and therefore had a cadre of scientists and researchers who could come forward with innovative ideas to confront the crisis. While Cuba has only two percent of the population of Latin America, it has almost 11 percent of the scientists (Rosset and Benjamin 1994).

Alternative Technologies

In response to this crisis Cubans and their government rushed to develop and implement alternatives. Because of the drastically reduced availability of chemical inputs, the state hurried to replace them with locally produced, and in most cases biological, substitutes. This has meant biopesticides (microbial products) and natural enemies to combat insect pests, resistant plant varieties, crop rotations and microbial antagonists to combat plant pathogens, and better rotations and cover cropping to suppress weeds. Scarce synthetic fertilizers were supplemented by biofertilizers, earthworms, compost, other organic fertilizers, animal and green manures, and the integration of grazing animals.

In place of tractors, for which fuel, tires, and spare parts were often unavailable, there was a sweeping return to animal traction (Rosset and Benjamin 1994).

When the crisis began, yields fell drastically throughout the country. But production levels for domestically-consumed food crops began to rise shortly thereafter, especially on Agricultural Production Cooperatives (CPAs) and on the farms of individual small holders or *campesinos.* It really was not all that difficult for the small farm sector to effectively produce with fewer inputs. After all, today's small farmers are the descendants of generations of small farmers, with long family and community traditions of low-input production. They basically did two things: remembered the old techniques—like intercropping and manuring—that their parents and grandparents had used before the advent of modern chemicals, and simultaneously incorporated new biopesticides and biofertilizers into their production practices (Rosset 1997b; 1997c).

The state sector, on the other hand, faced the incompatibility of large monocultural tracts with low-input technology. Scale effects are very different for conventional chemical management and for low external input alternatives. Under conventional systems, a single technician can manage several thousand hectares on a "recipe" basis by simply writing out instructions for a particular fertilizer formula or pesticide to be applied with machinery on the entire area. Not so for agroecological farming. Whoever manages the farm must be intimately familiar with the ecological heterogeneity of each individual patch of soil. The farmer must know, for example, where organic matter needs to be added, and where pest and natural enemy refuges and entry points are (Altieri 1996). This partially explains the difficulty of the state sector to raise yields with alternative inputs. A partial response was obtained with a program that began before the Special Period, called *Vinculando el Hombre con la Tierra,* which sought to more closely link state farm workers to particular pieces of land, but it wasn't enough (Enriquez 1994).

In September 1993 Cuba began radically reorganizing the state sector in order to create the small-scale management units that seemed most effective in the Special Period. The government issued a decree terminating the existence of the majority of state farms, turning them into Basic Units of Cooperative Production (UBPCs), a form of worker-owned enterprise or cooperative. Much of the 80 percent of all farmland that was once held by the state, including sugarcane plantations, was essentially turned over to the workers. The UBPCs allow collectives of workers to lease state farmlands rent free, in perpetuity. Property rights remain in the hands of the state, and the UBPCs must still meet production quotas for their key crops, but the collectives are owners of what they produce. What food crops they produce in excess of their quotas could be freely sold at newly opened farmers markets. This last reform, made in 1994, offered a price incentive to farmers to make effective use of the new technologies (Rosset 1997a).

The pace of consolidation of the UBPCs has varied greatly in their first years of life. With a variety of internal management schemes, in almost all cases the effective size of the management unit has been drastically reduced. It is clear that the process of turning farm workers into farmers will take some time—it simply cannot be accomplished overnight—and many UBPCs are struggling, while others are very successful. On the average, small farmers and CPAs probably still obtain higher levels of productivity than do most UBPCs, and do so in ways that are more ecologically sound.

Food Shortage Overcome

By the latter part of the 1990s the acute food shortage was a thing of the past, though sporadic shortages of specific items remained a problem, and food costs for the population had increased significantly. The shortage was largely overcome through domestic production increases which came primarily from small farms, and in the case of eggs and pork, from booming backyard production (Rosset 1998). The proliferation of urban farmers who produce fresh produce has also been extremely important to the Cuban food supply (GNAU 2000; Murphy 1999). The earlier food shortages and the resultant increase in food prices suddenly turned urban agriculture into a very profitable activity for Cubans, and once the government threw its full support behind a nascent urban gardening movement, it exploded to near epic proportions. Formerly vacant lots and backyards in all Cuban cities now sport food crops and farm animals, and fresh produce is sold from stands throughout urban areas at prices substantially below those prevailing in the farmers' markets. There can be no doubt that urban farming, relying almost exclusively on organic techniques, has played a key role in assuring the food security of Cuban families over the past two to three years.

An Alternative Paradigm?

To what extent can we see the outlines of an alternative food system paradigm in this Cuban experience? Or is Cuba just such a unique case in every way that we cannot generalize its experiences into lessons for other countries? The first thing to point out is that contemporary Cuba turned conventional wisdom completely on its head. We are told that small countries cannot feed themselves; that they need imports to cover the deficiency of their local agriculture. Yet Cuba has taken enormous strides toward self-reliance since it lost its key trade relations. We hear that a country can't feed its people without synthetic farm chemicals, yet Cuba is virtually doing so. We are told that we need the efficiency of large-scale corporate or state farms in order to produce enough food, yet we find small farmers and gardeners in the vanguard of Cuba's recovery from a food crisis. In fact, in the absence of subsidized machines and imported chemicals, small farms are more efficient than very large production units. We hear time and again that international food aid is the answer to food shortages—yet Cuba has found an alternative in local production.

Abstracting from that experience, the elements of an alternative paradigm might therefore be:

AGROECOLOGICAL TECHNOLOGY INSTEAD OF CHEMICALS: Cuba has used intercropping, locally produced biopesticides, compost, and other alternatives to synthetic pesticides and fertilizers.

FAIR PRICES FOR FARMERS: Cuban farmers stepped up production in response to higher crop prices. Farmers everywhere lack incentive to produce when prices are kept artificially low, as they often are. Yet when given an incentive, they produce, regardless of the conditions under which that production must take place.

REDISTRIBUTION OF LAND: Small farmers and gardeners have been the most productive of Cuban producers under low-input conditions. Indeed, smaller farms worldwide produce much more per unit area than do large farms (Rosset 1999). In Cuba redistribution was relatively easy to accomplish because the major part of the land reform had already occurred, in the sense that there were no landlords to resist further change.

GREATER EMPHASIS ON LOCAL PRODUCTION: People should not have to depend on the vagaries of prices in the world economy, long distance transportation, and superpower "goodwill" for their next meal. Locally and regionally produced food offers greater security, as well as synergistic linkages to promote local economic development. Furthermore such production is more ecologically sound, as the energy spent on international transport is wasteful and environmentally unsustainable. By promoting urban farming, cities and their surrounding areas can be made virtually self-sufficient in perishable foods, be beautified, and have greater employment opportunities. Cuba gives us a hint of the underexploited potential of urban farming.

The Cuban experience illustrates that we can feed a nation's population well with an alternative model based on appropriate ecological technology, and in doing so we can become more self-reliant in food production. Farmers must receive higher returns for their produce, and when they do they will be encouraged to produce. Expensive chemical inputs—most of which are unnecessary—can be largely dispensed with. The important lessons from Cuba that we can apply elsewhere, then, are agroecology, fair prices, land reform, and local production, including urban agriculture.

For those interested in learning from the details of the Cuban experience in resisting not only the United States but also the negative impacts of conventional industrial agriculture, we are proud to copublish this volume of essays by Cuban experts. Taken together this is a collection full of lessons and important experiences that will be useful in many settings and in many countries. It is with great pleasure and pride that we make it available to our readers.

References

Altieri, M. 1996. *Agroecology: The Science of Sustainable Agriculture.* Boulder: Westview Press.

Enriquez, L. 1994. *The Question of Food Security in Cuban Socialism.* Berkeley: Institute of International and Area Studies, University of California at Berkeley.

GNAU. 2000. *Manual Técnico de Organopónicos y Huertos Intensivos.* Havana: ACTAF, INIFAT, MINAG.

Lappé, F.M., and J. Collins. 1977. *Food First: Beyond the Myth of Scarcity.* New York: Ballantine Books.

Lappé, F.M., J. Collins and P.M. Rosset, with L. Esparza. 1998. *World Hunger: Twelve Myths,* second edition. New York: Grove Press.

Murphy, C. 1999. *Cultivating Havana: Urban Agriculture and Food Security in the Years of Crisis. Food First Development Report* No. 12. Oakland: Institute for Food and Development Policy.

Rosset, P.M. 1997a. La crisis de la agricultura convencional, la sustitución de insumos, y el enfoque agroecológico. (Chile) *Agroecología y Desarrollo* 11/12:2–12.

Rosset, P.M. 1997b. Alternative agriculture and crisis in Cuba. *Technology and Society* 16(2):19–25.

Rosset, P.M. 1997c. Cuba: Ethics, biological control, and crisis. *Agriculture and Human Values* 14:291–302.

Rosset, P.M. 1998. Alternative agriculture works: The case of Cuba. *Monthly Review* 50:3.

Rosset, P.M. 1999. *The Multiple Functions and Benefits of Small Farm Agriculture in the Context of Global Trade Negotiations. Food First Policy Brief* No. 4. Oakland: Institute for Food and Development Policy.

Rosset, P.M., and M. Altieri. 1994. Agricultura en Cuba: una experiencia nacional en conversión orgánica. (Chile) *Agroecología y Desarrollo* 7/8.

Rosset, P.M. and M. Altieri. 1997. Agroecology versus input substitution: A fundamental contradiction of sustainable agriculture. *Society & Natural Resources* 10(3):283–295.

Rosset, P.M. and M. Benjamin (eds.). 1994. *The Greening of the Revolution: Cuba's Experiment with Organic Agriculture.* Australia: Ocean Press.

PART I
History, Context, and Structures of Change

CHAPTER ONE

The Organic Farming Movement in Cuba

Fernando Funes, Pasture and Forage Research Institute

In recent decades organic farming has been growing in importance, and today it is recognized as a strong international movement. The main purpose of this movement is the search for an alternative to the conventional "Green Revolution" model of agriculture. Initially the Green Revolution had great impact on agricultural yields, but soon it revealed its fragility, vulnerability, and riskiness for the environment, human health, and agroecosystems, and led to little socioeconomic security for the poorest farmers. World recognition of today's environmental problems, which have caused and continue to generate global effects with unpredictable consequences, came together at the 1992 UN Earth Summit in Brazil. There is now great concern around the world for the environmental problems resulting from the industrial agriculture model, which have included erosion, salinity, and infertility in a large portion of our agricultural soils, the loss of biodiversity, growing deforestation, and socioeconomic problems in rural regions, including mass migration to cities.

Many alarms have been sounded and sent to the international scientific community echoing these problems. Three books voicing the concerns of the last two generations are *Silent Spring* (Carson 1962), *Our Stolen Future* (Colborn et al. 1996) and *World Hunger: Twelve Myths* (Lappé et al. 1998). These books examine the root causes of our problems, and use real examples to show the environmental consequences generated by the dominant agricultural model. They provide us with alternative pathways to redirect agricultural development in a more ecological direction.

In Cuba there were important historical precursors to today's organic farming movement. They range from our great agricultural thinkers of the past like Álvaro Reynoso, Francisco de Frías, Tranquilino Sandalio de Noda, Francisco

Javier Balmaseda, Antonio Bachiller y Morales, José Comallonga, Isaac del Corral, José L. Amargós, Juan Tomás Roig, Julián Acuña, Jesús Cañizares, among others, to the contemporary analysts referred to in this study. The early pioneers wrote from a naturalist background, and established an ecological agricultural tradition, providing future generations with important knowledge, concepts, and ideas through their valuable works.

In recent decades, especially in the 1990s, the organic movement in our country has grown, and great advances have been achieved through its practical application, not only in our agricultural systems, but also in our nation's conception of environmentally-based development. The Cuban state has been supportive, through the establishment of specialized institutions, legislation, research, teaching and extension, and through productive practice. According to Lane (1997), today's Cuban development model has the potential to transform our country into one of the first sustainable societies of the twenty-first century. She singles out our scientific and educational development as one of the basic pillars of this potentiality.

In this chapter I lay out the principles and progress of the Cuban organic farming movement, and provide background information for the topics that will be discussed in greater depth by the other authors of this volume.

Principle Characteristics of Cuba

The Republic of Cuba is an archipelago in the Caribbean Sea at the gateway to the Gulf of Mexico, and is very close to the Tropic of Cancer, which borders it on the north. Cuba has a surface area of 110,860 km^2 and measures 1,200 km in length. The country has a tropical climate characterized by abundant rainfall from May to October, with 80 percent of the 1,300 mm annual average rainfall falling at this time, and sparse and irregular rainfall during the dry season (November to April). The annual mean temperature is 25°C, ranging during the year from 23 to 27°C, and the average relative humidity is 80 percent (IM 2000). The predominant soil types are oxisols and ultisols (68 percent), with some inseptisols (16 percent) and vertisols (16 percent).

There are 11,142,600 inhabitants of Cuba, with a density of a little more than 100 inhabitants/km^2. Seventy-four percent of the population is urban. The capital is the city of Havana, and the official language is Spanish. There are fourteen provinces and one special municipality (the Isle of Youth). The ethnic composition of the population is 66 percent Caucasian, 21.9 percent *mestizo*, 12 percent black, and 0.1 percent Asian.

Cuban Agriculture before 1990

BEFORE THE REVOLUTION

In order to understand the organic farming movement and sustainable agriculture in Cuba, one must begin with the history of Cuban agriculture. In the pre-Columbian period, indigenous people practiced a rudimentary agriculture that, together with hunting and fishing, ensured their food supply. The beginning of sedentary Cuban agriculture is generally placed in the year 1511, during the colonial conquest period when Diego Velázquez initiated the first distribution of land.

In 1536 the royal occupation of land began in the town of Sancti Spíritus, and land holdings were granted for country houses and farms. The land distribution process in the sixteenth, seventeenth, and eighteenth centuries gave place to the settling of a class of rural landowners, and led to the birth of the Cuban farmer. The introduction of large numbers of African slaves—estimates of their numbers exceed 650,000—which were brought to the island beginning in the late eighteenth century, allowed the birth and the expansion of the sugarcane industry, a key point in the development of Cuban agriculture (Moreno-Fraginals 1978; Bergard et al. 1995). The farming methods used by indigenous peoples, who were exterminated after a brief time, leaving practically no traces, were merged with those of African slaves and techniques brought from Europe and other parts of America by the Spanish. These hybrid techniques, adapted to the natural conditions of the island, created a sort of sustainable agriculture, characterized in many cases by the careful selection and appropriate use of soils, the programming of sowing dates matched to local climates, the planting of polycultures and the use of crop rotations, the application of natural soil amendments and organic fertilization, etc.

For over four centuries there was a trend toward increasing small and medium scale land holdings, especially in crop-based production, as livestock production was still characterized by large properties from the sixteenth century until the beginning of the seventeenth century, later giving way through population growth and subdivision into medium and small-sized diversified share-cropped farms (Nova 1997).

As long ago as 1862, Francisco de Frías y Jacob, the Count of Pozos Dulces, was quoted as saying that "intercropping and crop rotation in Cuba will reverse the rampant land degradation caused by ignorance and greed"—revealing early agroecological thinking. National historians like Ramiro Guerra and Fernando Ortiz have brilliantly reviewed the special features of this period.

At the beginning of the War of Independence against Spanish colonialism (1895–1898) there were 90,700 working farms, and by the end of the war, there were 60,711, averaging 58 hectares in size. Those with less than 13 hectares predominated, occupying approximately 50 percent of the agricultural area of the country. Large farms of more than 135 hectares were largely devoted to sugarcane and livestock production. Even on sugarcane estate some lands were used for cattle ranching, forest production, and food, fruit, and other crops (Nova 1997).

At the beginning of the twentieth century, with American intervention and the beginning of large capital flows into the island, a period of sugarcane expansion began, based on specialization on large holdings, or in some cases, with associated cattle ranching. This was accompanied by the reduction in number and area of smaller, more diversified farms. By 1934 there were only 38,130 medium and small farms left. Later, large-scale rice production followed in the footsteps of sugarcane in the spread of *latifundios* (large estates). These large mono-crop *latifundios* were largely operated by American companies using the principles of "conventional" or "modern" agriculture.

In 1958, 56 percent of the Cuban population still lived in the countryside, and faced acute social problems. In the rural sector just 9.4 percent of the landholders owned 73.3 percent of the land, while 85 percent of the farmers rented their land; there were more than four million hectares left uncultivated on large estates, while 200,000 families were landless. Rural areas were characterized by high levels of illiteracy and infant mortality, and by unsanitary conditions. Out of a total population of a little more than six million, 600,000 Cubans were unemployed and more than 500,000 were part-time agricultural laborers who only worked four months of the year. Over half of the best agricultural lands lay in the hands of foreign owners (Castro 1975).

AGRICULTURE DURING THE REVOLUTIONARY PERIOD

After the triumph of the Cuban Revolution in 1959, land was distributed to more than 200,000 peasant families through the Agrarian Reform Laws of 1959 and 1963, while 70 percent of the *latifundio* lands were passed over to state control. At the same time great efforts were initiated in the areas of education, culture, health, and economic development. Rural towns, highways, and roads were built, along with rural electrification, as health services, schools, universities, and scientific centers were created throughout the country.

The main objectives of the revolutionary agricultural sector were:

- to meet the food requirements of the population
- to generate export earnings
- to provide raw materials for industry
- to eradicate poverty and unsanitary conditions in the countryside

In the early days of the revolutionary period, agricultural diversification and a greater emphasis on a nature-friendly type of agriculture were advocated.

However, the notion of agricultural progress soon succumbed to the demands and tendencies of the period, with a shift toward conventional agriculture. The blueprint for this shift came from the global strategy of the industrialized countries, including the Socialist Bloc in Eastern Europe. Although there were marked improvements compared to conditions faced in earlier periods, the modern agricultural model eventually showed signs of economic, ecological, and social problems, such as:

- over-specialization, monocropping, and excessive intensification
- excessive dependence on external inputs (fertilizers, pesticides, animal feed concentrates, farm machinery, and irrigation equipment)
- large-scale deforestation
- salinization, erosion, compaction, and fertility loss of soils
- unsustainable intensive factory farming systems of cattle, poultry, and pig production
- heavy rural-urban migration (in 1956, 56 percent of the population was rural, 28 percent in 1989, and less than 20 percent by the mid-1990s)

Nevertheless, the powerful social transformations brought by the Revolution prevented thousands of farmers from being driven into poverty by the bankruptcies that normally plague input-squandering conventional agriculture, contrary to the experiences of other underdeveloped or developing countries.

In this period, 80 percent of the lands were in the state sector, and 20 percent in the private sector. On the average, 1,300,000 tons of chemical fertilizers and 600,000 tons of feed concentrates for livestock production were used every year, together with $80 million worth of pesticides. The number of tractors employed in Cuban agriculture grew to 90,000, as the number of ox teams shrank to less than 100,000. In the first three decades of this period, the favorable terms of trade Cuba received from socialist countries, especially the Soviet Union, made such heavy investment in this agricultural model possible (Funes et al. 1999).

During this period small farmers from across the country formed the National Association of Small Farmers (ANAP), and many joined Agricultural Production Cooperatives (CPAs) or Credit and Service Cooperatives (CCS), while maintaining crop diversification and integrated farming practices on their lands. Valuable farming traditions survived as small farmers continued to use animal traction and intuitively practiced agroecological sciences, which kept the management and economics of their farms on a sustainable basis, using very few or no external inputs.

Recent Changes in Cuban Agriculture

In the 1970s the Cuban government became aware of some of the problems of the dominant agricultural model, and began to introduce changes that would lead to an agricultural system using fewer inputs and that was more rational and in keeping with the current situation. Policies were initiated to substitute for imported inputs and raw materials, and to stimulate financial and material savings in all sectors, and a new emphasis was placed on profitability and self-sufficiency.

In addition, research centers changed their objectives and strategies to focus on programs with a more rational and sustainable basis. In this period the world energy crisis and the rising prices of imported fuels, fertilizers, feed concentrates, pesticides, and other manufactured products had a negative impact on Cuba's agricultural economy. The 1980s saw growing research, development, and extension of methods to substitute for imported inputs (Funes 1997).

The solid research system that was founded in the 1960s was strengthened, in the Ministry of Agriculture (MINAG), with 17 research centers and 38 experimental stations (795 researchers and 168 Ph.D.s in different fields), in the Ministry of Higher Education (MES) with its network of research centers and universities, in the Ministry of Education (MINED), and in the other institutions that carry out agroecological research today.

The Crisis of the European Socialist Bloc: The "Special Period" in Cuba

In 1989 an acute crisis began suddenly with the collapse of the European socialist countries and the disintegration of the Soviet Union. We must keep in mind that in any event Cuba is not blessed with abundant capital nor with sufficient domestic energy supplies. Prior to 1989 more than 85 percent of our trade was with socialist countries in Europe, and a little more than 10 percent with capitalist countries. Cuba imported two thirds of its foodstuffs, almost all of its fuel and 80 percent of its machinery and spare parts from socialist countries. With the crisis, Cuba's purchasing capacity was reduced to 40 percent, fuel importation to a third, fertilizers to 25 percent, pesticides to 40 percent, animal feed concentrates to 30 percent; and all agricultural activities were seriously affected. Suddenly, $8 billion a year disappeared from Cuban trade. Between 1989 and 1993, the Cuban GNP fell from $19.3 to $10.0 billion. Imports were reduced by 75 percent, including most foodstuffs, spare parts, agrochemicals, and industrial equipment. Many industries were forced to close, and public transportation and electric plants worked at minimum capacity (Espinosa 1997). Unexpectedly, a "modern" and industrialized agricultural system had to face the challenge to increase food production while maintaining production for export, all with a more than 50 percent drop in the availability of inputs.

To face the crisis the Cuban government put economic austerity measures and emergency changes into practice, such as: a new domestic economic policy, an opening to foreign investment, the liberalization of the rules governing the possession of dollars by Cuban citizens, and the granting of licenses for private work in various sectors. Together with structural reorganization, new agricultural techniques developed in recent decades received their first extensive implementation, as a variety of measures were introduced, including:

- decentralization of the state farm sector through new organizational forms and production structures
- land distribution to encourage production of different crops in various regions of the country
- reduction of specialization in agricultural production
- production of biological pest controls and biofertilizers
- renewed use of animal traction
- promotion of urban, family, and community gardening movements
- opening of farmers' markets under "supply and demand" conditions

The objective of agrarian policy during this "Special Period" was to move to a low external input form of agriculture, while at the same time boosting production. This required a greater level of organization of Cuban research and agricultural extension structures, a better flow of information, and a reduced emphasis on technologies requiring a lot of capital and/or energy.

Simultaneously, the United States tightened the economic blockade of Cuba. In 1992 the Torricelli bill was approved, barring shipments to Cuba of food and medical supplies by overseas subsidiaries of US companies, and later the Helms-Burton Act (1996) restricted foreign investment in Cuba. These laws have been strengthened by a variety of amendments, multiplying the effects of the blockade, which took on increasingly cruel and extreme characteristics.

In spite of everything, Cuba has managed to maintain high scores on most social indicators. The literacy rate in Cuba is still above 95 percent, education is free and compulsory until age 16, the average level of schooling is the ninth grade, and of 11 million inhabitants, more than half a million are university graduates. Infant mortality is 6.4/1000 live births. There are approximately 60 doctors for every 10,000 inhabitants and average life expectancy is still above 75 years. For every 100 economically active Cubans, 96 have jobs (Francisco 2000). Women make up 43 percent of the workforce, employed principally in education, health care, and in the scientific and technical sectors. While Cuba accounts for only two percent of the Latin American population, it has twelve percent of its scientific workers (Ellwood 1998).

Many sectors of agricultural production have been recovering in recent years, especially roots and tubers and fresh vegetables (which reached historic production levels in 1999), and national forest cover, which currently surpasses 21 percent of the national land area (Peláez 2000). Urban agriculture has made an important contribution to food security. Significant progress has been achieved with programs for small-scale rice and medicinal plant production, the use of animal traction has been rediscovered, farmers' markets are growing rapidly, and MINAG and other state entities have opened other "fixed price ceiling" marketplaces which offer affordable food prices.

Administrative Structures for Agriculture

The Ministry of Agriculture is the national organization with the overall charge of directing and regulating agricultural and forest production to meet the food needs of the population, raw material needs of industry, and tourism requirements, as well as to substitute imports and to encourage exports with maximum efficiency. To achieve these objectives, it relies on rational use of farmland, water, and technical supplies, and on efforts geared toward conserving the soil and the genetic stock of domestic and wild fauna, and the preservation and use of resistant crop varieties and seeds in non-sugarcane agricultural and forest production. MINAG also guarantees services and inspections for animal and plant health, and environmental protection, and safe workplace conditions (MINAG 1999). Its total workforce consists of 1,153,000 employees (including Basic Units of Cooperative Production workers), of which 26,352 are professionals with higher degrees and 62,200 are technicians (MINAG 1999). MINAG's State Enterprise System consists of 487 enterprises, and 222 independent farms, with some 400,000 workers in agricultural (69 percent), silvicultural (10 percent), industrial (9 percent), construction (2 percent), transport (1 percent), and commerce (9 percent) sectors (MINAG 1999).

Given the overriding importance of sugarcane and the sugar industry to the Cuban economy, the Ministry of Sugar (MINAZ) is responsible for an agricultural area of approximately 1,500,000 hectares, with functions similar to those of MINAG in the agricultural and agro-industrial sectors.

Other organizations closely related to agriculture include:

The National Association of Small Farmers (ANAP) provides organizational and productive support, as well as training, promotion, marketing, international cooperation, for small farmers, whether they are members of CPAs,

CCSs, or individual farmers. It has helped its members preserve a large portion of Cuba's farming traditions, experiences, and culture, which have been and continue to be of great importance for the shift toward sustainable and agroecological agriculture.

The Ministry of Education (MINED) is entrusted with technical education in the countryside, relying on a network of Agricultural Polytechnic Institutes (IPAs, essentially rural vocational high schools). These institutes have agricultural production areas that are looked after by their own students, providing both a theoretical and practical education, while at the same time providing food for the students and professors.

The Ministry of Higher Education (MES) is responsible for university and post-graduate teaching. All agricultural universities are included in its structure, as well as several research institutes and experiment stations, some of them of great national and international prestige, that give important support to the activities of MINAG.

The Ministry of Science, Technology, and the Environment (CITMA), created in 1994 during the Special Period, creates and implements state policy concerning scientific, technical, and environmental problems, of which the agricultural sector has a high priority.

Other ministries such as the Ministry of Food (MINAL), the Ministry for Foreign Investment and Economic Collaboration (MINVEC), and institutions such as the National Institute of Hydraulic Resources (INRH) all have close links with the agricultural sector.

Changes in the Structure of Production

In September 1993 the Cuban state made changes to the landholding organization of the former state farms, a large part of which have now been turned into Basic Units for Cooperative Production (UBPC). These lands were granted under free usufruct basis (no rent is paid) to those that previously had worked for the state farms, who then also purchased the associated means of production (farm equipment and implements, farm animals, buildings, etc). The UBPCs generally cover larger areas than the CPAs, and are self-administered and financially independent.

Today the private sector occupies a central place in agricultural production, as it includes the CPAs and the CCSs (created in the 1970s and organized under ANAP), the former state farms that were transferred to the UBPCs, as well as other private producers. Moreover, the so-called New Type State Farms

(GENTs) were created in those places where conditions were unfavorable to forming UBPCs, and whose main feature is having greater autonomy of management compared to the previous state farms. The amendment of MINAG's Regulation 419 gave out 80,748 hectares of land to new farmers to promote coffee production, 53,948 hectares for tobacco, and 17,004 hectares from this new distribution of land (MINAG 1999).

Land Tenure Structures

STRUCTURE	ORIGIN	TENURE	BENEFITS
CPA	Individual landowners	Voluntary donation of land to form cooperatives and associations	Wages, based on effort and participation
CCS	Former tenants, farm workers, partisans, sharecroppers, and small landowners	Private lands and in-usufruct lands	Bank credits, profit sharing
UBPC	Former state farm workers	Collective usufruct of lands with purchased means of production, animals, etc.	Wages, based on effort and participation
Lands in usufruct, rural sector	State lands. Mainly in coffee, cocoa, and tobacco	Usufruct of state lands	Sale to the state of the principle crop, family subsistence production, and free market sale of surplus crops
Urban agriculture	Backyards, roofs, balconies, and urban and peri-urban plots	Private or usufruct up to 0.25 hectares	Family self-provisioning, neighborhood sales of vegetables, flowers, herbs, and animals
GENT	State farms lacking the conditions needed to form UBPCs	State lands, state means of production, with greater administrative autonomy	Wages, by type of work and production results
State enterprises	State lands	All means of production belong to the state	Salaried workers, food and export production

The Organic Farming Movement in Cuba

The extraordinary work of the movement's precursors mentioned earlier, and subsequent generations that together with thousands of peasants and farmers imparted their knowledge, ideas, and experiences, have forged today's

Cuban organic farming movement and its basic principles. In the 1970s and 1980s many Cuban scientists and farmers started searching for alternatives to high input agriculture. Several research centers adopted this line of work, and a consciousness was gradually created that Cuba could reduce the use of imported inputs, making agricultural systems more sustainable from economic and environmental viewpoints.

In light of the historical background and the recent transformations initiated by the Cuban state, in 1992 a group of professors and researchers, mainly from the Ministry of Higher Education, aware of the need to promote alternatives to conventional agriculture, joined together at the Agrarian University of Havana (UNAH—at that time called the Advanced Institute of Agricultural Sciences of Havana, or ISCAH) to discuss agroecological ideas. They organized the First National Conference on Organic Agriculture, held in May 1993 at the National Institute of Agricultural Sciences (INCA), with the participation of more than 100 Cuban delegates and 40 from abroad, and founded the Formative Group of the Cuban Organic Farming Association (ACAO). Its principle objectives were:

- to develop a national consciousness of the need for an agricultural system in harmony with both humans and nature, while producing sufficient, affordable, and healthy food in an economically viable manner
- to develop local agroecological projects, and promote the education and training of the people involved in this new paradigm of rural development
- to stimulate agroecological research and teaching, and the recovery of the principles on which traditional production systems have been based
- to coordinate technical assistance to farmers and promote the establishment of organic and natural agricultural production systems
- to encourage the exchange of experiences with foreign organizations (with emphasis on the Latin American tropics and subtropics), and with specialists in sustainable agriculture and rural development
- to promote and publicize the importance of marketing organic products

Since the founding of ACAO a strong effort has grown through workshops, field days, talks in universities and research centers, conferences, events, and meetings with farmers, as well as portable agroecological libraries which rotate through different production, research, and education centers, agricultural cooperatives, and other interested institutions.

Simultaneously, a program was initiated with farmers to create "Agroecological Lighthouses," or farms where agroecological concepts are applied and which promote sustainable production systems in different regions of the country. The lighthouse project began on three CPAs in two municipalities of the province of Havana, where the area of influence covered some 150 families and approximately 400 farmers. The Sustainable Agriculture Networking and Extension (SANE) project of the United Nations Development Program (UNDP) supported this program, which also included training at the UNAH and at Agricultural Polytechnic Institutes (IPAs). Seven new lighthouses have recently been initiated on UBPCs, CCSs, and CPAs in the provinces of Pinar del Río, Havana, City of Havana, Cienfuegos, Sancti Spíritus, Villa Clara, and Las Tunas.

Agroecological farm integrating agriculture and cattle ranching. Pasture and Forage Research Institute, Havana.

Projects like the lighthouses have been carried out with the financial support of Bread for the World (Germany), HIVOS (Netherlands), and Oxfam America (United States). At the national school for farmers of ANAP, conferences, participatory meetings, and discussions of videos have been offered on theoretical and practical topics. Finally, many courses and workshops on agroecology and organic agriculture have been offered in different provinces of the country.

The Formative Group of ACAO worked together over several years with different entities including MINAG, MINAZ, MINED, MES, and CITMA, and had strong links to ANAP and the Council of Churches of Cuba (CIC). Work with these institutions was aimed at the production and marketing of organic products, with collaboration and interaction in the creation of educational, research, and development programs on sustainable agriculture.

In 1995 and in 1997 the Second and Third National Conferences on Organic Agriculture were held at the Institute of Animal Science in the Province of Havana and at the Central University of Villa Clara, respectively. In the second conference, in coordination with participant institutions, a variety of activities were organized including a study tour of different regions with organic farming activists and experts from diverse countries, especially from Latin America, two specialized workshops (biological pest control and soil tillage with animal traction), two mini-courses (intercropping and design of agroecological systems), a three-day conference, and the Second International Course on Organic Farming.

Plenary session of the Second National Meeting for Organic Agriculture, Animal Science Institute. Havana, 1995.

At the third conference, more than 400 delegates (180 foreigners and 240 Cubans) participated for three days. Prior to the meeting workshops were held on development in mountain regions, urban agriculture, and ecological pest management. The International Meeting of the SANE Project was held concurrently, and in the immediate aftermath a meeting of the Pesticide Action

Network (PAN) and the Third International Course on Organic Farming were held. In the context of the third conference, the First Latin American Conference of Organic Certifiers was held, to learn about certification initiatives, and to develop a common work plan in search of alternatives to solve the most relevant problems of the region.

In the educational realm, the Center for the Study of Sustainable Agriculture (CEAS) at the Agrarian University of Havana has initiated master's and Ph.D. programs in agroecology, and since 1997 has offered an annual correspondence course on agroecology with a high level of participation throughout the country.

Courses on certification of organic products and on the production and certification of organic coffee have been offered, as well as various conferences and workshops for farmers. The Italian Association for Organic Agriculture (AIAB) has supported this effort by training two professionals in Italy as international organic inspectors, recognized by the European Economic Union.

The magazine *Agricultura Orgánica (Organic Farming)* commenced publication three times a year in 1995. Its objective has been to analyze, debate, and disseminate different aspects of and advances in organic agriculture, as well as news of low external input and agroecological technologies, with an eye towards system sustainability. Environmental, social, and economic topics are also discussed, especially in light of the problems provoked by conventional agriculture in different parts of the world. A broad national and international readership indicates the high level of acceptance of this magazine (Monzote and Funes 1997).

There has been strong national and international exchange and cooperation. Practically all members of the Formative Group of the ACAO come from or work in centers or institutions where they develop programs for, or have a links with, this form of agriculture, often working together in projects, courses, publications, and in the organization of meetings. At the international level, in addition to collaborative work, active participation in international meetings and study tours throughout the country for foreign delegations have resulted in thousands participating in making Cuba's experiences in organic and sustainable agriculture more widely known. Cuban delegations of specialists and farmers have reported on our agricultural practices in Bolivia, Colombia, Peru, Venezuela, Guatemala, Nicaragua, Costa Rica, Mexico, Haiti, Spain, the Netherlands, Australia, New Zealand, Laos, Malaysia, Nepal, Sri Lanka, the United States, and other countries, with exchanges with thousands of people in these regions.

In 1996 the Formative Group of the ACAO received the Saard Mallinkrodt Award (which was announced at the International Federation of Organic Agriculture Movements meeting in Copenhagen, Denmark) for its work promoting organic agriculture. In April 1999, ACAO changed its name upon joining, and becoming a section of, the Cuban Association of Agricultural and

Forestry Technicians (ACTAF), where it is now called the Organic Farming Group (GAO), achieving officially recognized status in the country. GAO continues to work under the same premises of the promotion and development of agroecological agriculture.

In December 1999 GAO was awarded the "Alternative Nobel Prize," or Right Livelihood Award, in a solemn session of the Swedish Parliament, for its work disseminating and promoting organic agriculture. This came in recognition of all of GAO's fine work in organic farming, though it is also a testimony to the thousands of Cuban men and women who are developing and implementing organic practices day by day, as another way of resisting the tightening of the economic blockade that our country has suffered for more than 40 years.

Scientific, Technical, and Socioeconomic Bases for the Development of Organic Agriculture in Cuba

From the beginning of the "Special Period" the repercussions of the crisis were felt throughout the country. However, there was an immediate response, and in agricultural activity it was essential to rely on the alternatives that different research centers had been experimenting with for a number of years, as well as the possibility of recapturing the experiences of *campesino* farmers who had knowledge that had been passed down from former generations, but that had been "forgotten" or displaced by conventional agriculture. The accumulated cultural, political, and technical preparation of the Cuban people, built up throughout the revolutionary period, played an essential role in helping Cuba face the abrupt change.

Thus, the Ministry of Agriculture rapidly started to apply initial research results on a large scale, in order to reduce and in some cases to offset the effects of the crisis on agriculture. Throughout the country, results obtained in decades prior to the 1990s were applied. Other ministries quickly took legal, economic, and social measures, trying to adapt to the new conditions.

Soon alternatives became realities, and a consciousness was gradually created among many farmers, technicians, researchers, professors, and officials, that it was possible to create an agriculture with a different vision. People came to believe that productive harvests could be obtained on positive cost-benefit terms, while protecting the environment and nature, without polluting soils, water, and air, yet producing healthy foods without excessive energy use, and with reduced capital investment. After the beginning of the Special Period there was a general shift in the directions of research, education, and production. In this latter stage, the performance of these "new techniques" has been gradually shown to be effective in solving many critical agricultural problems. These scientific and technical activities as well as the accumulated experience of our farmers have had decisive results (Funes et al. 1999).

We should admit at this point that the principal techniques receiving widespread application have only been of the "input substitution" or "horizontal conversion" varieties (reduced input use, soil recovery techniques, etc.). A narrow technical focus has not yet allowed us to take significant advantage of the mechanisms of synergy that would be made possible in a more completely agroecological conception of agricultural development. Nevertheless, this first phase has been, and continues to be, extremely important in solving today's set of challenges—and has provided the basis for the widespread consolidation of organic farming over a large scale. Here I briefly highlight some of the areas in which important results have been obtained in recent years, which other authors in this volume will analyze in greater detail.

Organic Fertilization and Soil Conservation

We have many results on the use of manure, sugarcane filter-cake mud (*cachaza*), organic fertilizers, compost, bioearth, worm humus, residues from sugarcane collection centers, waste waters, cover crops, mulch, biofertilizers (*Rhizobium, Bradyrhizobium, Azotobacter, Azospirillum,* phosphorus solubilizing microorganisms, vesicular-arbuscular mycorrhizae), and other materials. In general these organic fertilizers and biofertilizers produce higher yields and improve soil cover, dry matter content, and soil properties; and have allowed substitution of organic methods for chemical fertilizers to meet the nutrient requirements of crops previously met through external inputs. In some cases, crop water requirements have been reduced as well.

Different institutions have worked on these topics, including the National Institute for Fundamental Research on Tropical Agriculture (INIFAT), the Liliana Dimitrova Horticultural Research Institute (IIHLD), the National Research Institute for Tropical Roots and Tubers (INIVIT), the National Institute of Agricultural Sciences (INCA), the Soil Research Institute (IIS), the Pasture and Forage Research Institute (IIPF), the Animal Science Institute (ICA), the Institute of Ecology and Systematics (IES), the Central University of Las Villas and other universities, and the Agricultural Polytechnic Institutes (IPAs), as well as other research centers.

Ecological Management of Pests, Diseases, and Weeds

One of the main challenges in the conversion to organic agriculture is the elimination of pesticide use. The work accomplished by the National Plant Protection Institute (INISAV) of MINAG, with the creation of a national network of Centers for the Production of Entomophages and Entomopathogens (CREEs), where artisanal and decentralized production of biocontrol agents is carried out, demonstrates that it is possible to count on local ecological solutions to combat pests and diseases. These practices are critical to the input substitution phase of organic agriculture, and are a world-class example of massive application. As such they have captured the attention of foreign scientists and organic farmers. Intense use is made of biological control by using predators, insect pathogens, and disease antagonists to control agricultural pests and diseases. Plants with insecticidal, fungicidal, bactericidal, and herbicidal qualities are used as well, as are parasitic nematodes, among others.

Two hundred and seventy-six CREEs have been established, of which MINAG and MINAZ are responsible for 222 and 54, respectively. Three industrial-scale plants are functioning as well, and another is being built by MINAG. All of them provide services to farmers (Pérez 1997). Presently, some one million hectares are protected by the application of biological controls, out of the five million total hectares devoted to agriculture in the country, covering a wide range of crops (INISAV 2000).

The research centers of MES also stand out in plant protection work, including the Agrarian University of Havana, as does the Reference Center for Biological Products of the MINAZ, some of the IPAs of the MINED, and various CREEs linked to ANAP, among others.

Crop and Livestock Management

CROP ROTATION AND POLYCULTURE: Both methods are commonly used in organic agriculture, and our work has shown positive responses, mainly related to land use and crop yields. Different polycultures or intercrops have been very good for improving soil coverage and quality, with land equivalent ratios (LERs) ranging from 1.01 to more than 3.0. Both have played key roles in controlling harmful pests and diseases (Hernández et al. 1998; Serrano 1998), and have helped guarantee the food supply to the population in recent years. While there are many research centers studying these topics, with important results, it has been the *campesinos* who never abandoned these practices who have made the greatest contributions.

LEGUME-BASED LIVESTOCK SYSTEMS, SILVO-PASTORAL SYSTEMS, AND INTEGRATED CROP-LIVESTOCK SYSTEMS: Livestock production has been one of the most affected sectors of agriculture, since in recent decades the development of cattle, poultry, and pigs was highly productive but based on specialized breeds, under intensive systems based on abundant consumption of imported feed concentrates. Animals raised under such conditions were highly vulnerable to changes in their diet and management, especially the drastic changes caused by the crisis. Productive and reproductive indices fell, especially in milk production. Obviously, the drop was greatest in the most "developed" zones of the country, and lower in those where less intensive systems still predominated.

Livestock systems take much longer to recover than crop production. Steps have been taken to return to locally adapted breeds and crosses for all livestock species, and free range rearing of poultry and pigs, using a natural, local resource diet, the use of legumes in protein banks and silvo-pastoral systems for cattle, along with diversification and greater integration of livestock with other farming activities (Muñoz 1997; Monzote and Funes-Monzote 1997). The research conducted at the Pasture and Forage Research Institute, the Animal Science Institute, and the "Indio Hatuey" Pasture and Forage Experiment Station have played a key role in these changes.

Ecological Soil Management

The National Service for the Use and Management of Agricultural Lands carries out soil studies ranging from detailed mapping to research on the needs and nutritional requirements of crops. The Soil Research Institute and the Agricultural Mechanization Research Institute of MINAG, as well as the Agricultural Mechanization Center of the UNAH, have developed alternatives for the management, conservation, and recovery of compacted, salinized, eroded, and other degraded soils. Today Cuban technical personnel have a sufficient body of knowledge regarding soil conservation and recovery. Organic techniques, such as the use of living barriers, ground cover with locally adapted pasture species, contour plowing, etc., are used in cases of severe degradation, though there is still much to be done (Durán 1998; Riverol 1998). In contemporary Cuba we have a conservation tillage system developed completely within the country based on scientific research and the accumulated knowledge and wisdom of Cuban *campesino* tradition.

Successful Organic Farming Experiences

URBAN AGRICULTURE: In the early 1990s, a strong urban agriculture was born, in which thousand of people produce food using organic methods that help supply basic foodstuffs to urban families. The variety of modalities developed include "organoponics," intensive vegetable gardens, plots, and backyards, suburban farms, self-provisioning by large enterprises, institutions, and offices, home gardening, etc. (Companioni et al. 1997). The effectiveness of organic techniques in urban gardening has been clearly demonstrated, and it is here that we are possibly closer to the ideal of integrated agroecological systems, due in part to the prohibition of the use of chemicals because of the proximity to dense human populations. The National Urban Agriculture Program has been led by INIFAT with practically all the research centers of the nation, although its main protagonists have been farmers. Their daily effort and that of their families have taken this movement forward. The program carries significant weight in national production of fresh produce and its consumption by the Cuban population.

POPULARIZATION OF SMALL RICE PRODUCTION: Widespread small-scale production of rice—a staple food in the Cuban diet—appeared spontaneously as a consequence of the economic crisis. In light of this movement, MINAG decided to organize and supply technical support to this sustainable production alternative with the Rice Research Institute playing a leading role. Today there is a "popular rice" production program, which consists of producing this cereal in small plots to ensure family consumption and the sale of surplus production—an important source of income for many families. This type of small-scale production has been a key factor in the stable presence, at acceptable prices, of rice in farmers' markets (Socorro et al. 1997).

MEDICINAL PLANTS: In 1992 organized production of medicinal plants began in Cuba, although there was a preexisting popular tradition regarding their cultivation and use. Today there is a growing trend toward rediscovering the use of "green" medicine for the prevention or cure of various diseases, and to cover the deficit of medicines during the Special Period. The program started as an initiative of the Ministry of Defense, but is now a joint effort of the Ministry of Public Health and MINAG. The latter is in charge of producing and drying the plants in bulk, and the former processes, packages, and distributes the medicines through pharmacies and hospitals. Currently there are thirteen provincial farms and 136 municipal modules with 700 hectares of medicinal plants under organic production. The current annual production of medicinal plants, and of herbs and plants used for dyes, is 1,000 tons, with plans to increase that amount in the coming years.

Incipient Organic Development Programs

ORGANIC SUGAR PRODUCTION: Recently the first experiments on organic sugar production have been carried out. The most advanced one is underway at a small sugarcane mill belonging to the Central University of Las Villas, which has started production on a pilot basis, in a joint venture with the firm Naturkost und Naturwaren from Germany, and with technical support, especially regarding certification, from the Italian Association for Organic Agriculture.

Initial steps are being taken to initiate organic sugar production at Agro-Industrial Complexes (CAI) of MINAZ; the first will be at the "Carlos Baliño" CAI in Villa Clara, which will manufacture approximately 3,000 tons of organic or ecological sugar on a commercial scale. According to current plans there will be at least one mill producing organic sugar in each province in the coming years (Varela-Pérez 2000). In these sugarcane projects, organic or ecological practices are used that include intercropping or rotation with soybeans and other legumes, biological pest control and the use of biofertilizers, compost, *cachaza* (sugar filter cake mud), and organic fertilizer.

ORGANIC FRUIT PRODUCTION: Many fruit tree plantations in Cuba—totaling about 32,000 hectares—have been grown without the use of agrochemicals during the 1990s. In these areas a new program has begun, to select plantations that demonstrate adequate conditions to restore their productive potential; and a shift towards organic practices has begun.

In the case of citrus, a production program for organic fruits and juices started selecting and converting commercial areas in 1997. At present there are citrus enterprises in organic transition in the provinces of Havana, Cienfuegos, Ciego de Ávila, Granma, Guantánamo, and the Isle of Youth (ACTAF 2000). In these provinces projects with foreign organizations are being developed with technical assistance from the Citrus and Fruit Research Institute and with the participation of the IIPF.

With the tourism boom, other tropical fruit trees are being produced under sustainable practices, and ecological farms are being promoted for direct consumption of these tropical fruits in tourism enterprises. More than five projects of this type are supported by the United Nations Food and Agriculture Organization (FAO) and various NGOs. Simultaneously, inroads are being made in the organic export market for coconut, and there are 300 hectares in the conversion process. The development of organic pineapple and mango is also planned, as well as the use of compost prepared from the solid wastes from citrus processing plants (five plants), enriched with sugar filter cake mud and biofertilizers (IICF 2000).

ORGANIC COFFEE AND COCOA: There is growing demand in the international market for organic coffee and cocoa. A project has been developed for the conversion of areas of these crops to organic. There are 3,000 hectares of coffee plantations selected in the western mountains of the provinces of Guantánamo and Santiago de Cuba. An estimated production of 150 tons will be exported to Europe. In the province of Guantánamo, 1,500 hectares of cocoa are in the conversion phase, with the possibility of producing approximately 200 tons in the first few years. At present the National Coffee and Cocoa Research Station (ENCC) works on the training of technical personnel and the implementation of organic practices (ENCC 2000).

Publications, Education, Training, and Research

The rapid movement towards a sustainable agriculture is being led by Cuban universities with the UNAH in the vanguard, developing courses and activities to train and update graduates in agroecological methods focusing on the substitution of knowledge- and management-intensive methods for high-input technologies. The Center for the Study of Sustainable Agriculture of the UNAH offers a comprehensive program including short courses, practical training, a correspondence diploma in agroecology and sustainable agriculture, a masters degree, and a doctorate in agroecology (García et al. 1999).

Thousands of *campesinos* have excelled as students of agroecology at the Niceto Pérez National Training Center of ANAP, in turn giving a shot in the arm to their respective cooperatives and organizations. ANAP's participatory farmer-to-farmer methodology has been very effective. Other outstanding results have been obtained in urban agriculture, and in joint activities of ANAP with the support of the Department of Project Coordination and Assistance of the Council of Churches of Cuba (Sánchez and Chirino 1999).

According to García (1999), other successes in agroecological training in Cuba have been achieved at research centers, IPAs, at the Cuban Animal Production Association (ACPA) by the GAO, and at the Agroecological Lighthouses. Participants have included professors and researchers at all levels; farmers are daily getting deeper into this type of agriculture and are excellent promoters and disseminators. To the thousands of people trained in formal courses, we can add hundreds of thousands of farmers trained by the MINAG, MINAZ, and ANAP, where participatory techniques and informal methods are used (Monzote 1999).

The Ministry of Science, Technology, and Environment has given priority in recent years to research in sustainable agriculture, through the approval of different research projects on crop diversification, agroecology, integrated crop-livestock production, organic and sustainable agriculture, etc. In many cases these projects have received additional support through financing from international organizations. Almost all the research centers of MINAG, MINAZ, MES (all also part of CITMA), along with universities, polytechnical schools, and others linked to agriculture, have been involved to a greater or lesser extent in these research programs with very important results.

In recent years, journals, books, pamphlets, fliers, press notes, radio and TV shows, and other activities have supported this new approach to Cuban agriculture. Some of these efforts include the publication of the *Agricultura Orgánica* magazine, the *Se Puede* newsletter of the Man and Nature Foundation's Permaculture Group, the television show "De Sol a Sol" from MINAG and the Cuban Institute of Radio and Television, and a number of radio shows that have broadcast the agroecological message to the people. Over many years the television show "Hoy Mismo" carried out excellent educational work in this field, and is remembered for the recently deceased Cuban organic activist, Professor Manuel Alvarez Pinto (Funes et al. 1999).

The collaboration and international aid received from different organizations in these initial stages of agroecological education, research, and publication includes support from the FAO; UNDP; the International Federation of Organic Agriculture Movements (IFOAM); the Latin American Consortium for Agroecology and Development (CLADES); Food First; AIAB; Bread for the World (Germany); the Latin American Agroecological Movement (MAELA); Oxfam; Humanist Institute for Development Cooperation (HIVOS); International Center for Rural and Agricultural Studies (CERAI); Organization for Costa Rican Development (CEDECO); and others. In exchange, Cuban professors, researchers, and farmers have shared their knowledge and experience on this subject in different countries, especially in Latin America, though also in other regions of the world.

The Present and Future Challenge

Organic farming and agroecology do not just represent a change of technological models, but of the very way in which we conceive of agriculture. This process automatically means a change in social consciousness, in tune with local reality.

Organic agriculture and agroecology make sense in the Cuban socioeconomic context, since as a rule this type of agriculture maintains a revolutionary worldview. Its principles run counter to the vicious globalization promoted by neo-liberalism, and are more in favor of a socially just and solidarious, more *human* globalization, without dependency on transnational corporations and in favor of self-sufficiency. Agroecology does not harm the environment, reduces the role of middlemen and intermediaries, develops the consciousness of farmers, and applies knowledge rather than crude technological recipes. It is an ally of nature and considers the farmer as a cultural and not just productive unit.

From a social viewpoint, Cuba is ideally situated to demonstrate the full possibilities of organic farming and to achieve truly sustainable agricultural systems (Monzote and Funes 1997). The favorable conditions present in Cuba include:

- a strong demand for agricultural products
- plenty of qualified personnel linked to agricultural activities
- a population experienced in community work
- administrative and social structures that support food self-sufficiency
- official mass media willing to sponsor publicity campaigns for the peoples' benefit
- research results that are compatible with the new model

- the return of many people to the countryside in recent years
- organizations dedicated to the creation of an agroecological culture

Organic farming is generally attained through a gradual conversion process, rather than through drastic changes and a sudden rupture with previous productive systems throughout a country, as occurred in Cuba due to the economic crisis. Yet due to these circumstances, Cuba has the conditions to continue perfecting agroecological production with fewer inputs, attuned to the specific conditions of each region, crop, productive purpose, and technological and economic conditions. The scene is also set to make rapid conversions in some commodities such as coffee, citrus and other fresh or processed fruits, sugar, and honey. Nevertheless, for strategic and practical reasons, some conventional agricultural systems will be maintained while the organic and agroecological systems are being fully developed.

The current economic conditions of scarce foreign exchange also favor implementation and marketing of organic agriculture, if the high cost of conventional agriculture is taken into account. Nevertheless, a product cannot be certified and sold as organic unless the established regulations for this type of agriculture have been followed. Thus, in the immediate future, we need to develop a policy in this regard, a process that has already begun. Concerns include definition of standards and legal regulations controlling organic production, a national certification system, market surveys to define priorities and possibilities inside and outside of the country, and a diagnostic process for each crop. In terms of a certification system for organic production, we must focus primarily on developing a national consciousness concerning the consumption of organic foods, and their permanent link with health and the environment, and how they help maintain our independence from transnational corporations.

While the first steps have been taken with the successful implementation of input substitution, the future challenge will be to develop more complex agroecological systems that will integrally and coherently combine crops, livestock production, forest management, and other subsystems, based on organic and sustainable methods geared toward taking full advantage of synergistic mechanisms. In this regard we have highly promising experimental results and the empirical experience of those farmers who have traditionally practiced agroecology on their farms, where animals, crops, fruit, and timber trees are fully integrated, while residues are recycled, and animal traction and wind energy are used (SANE 1999).

Research and development programs must continue to: demonstrate the vast potential of organic farming and agroecology; develop ever more effective methodologies for extension; increase the number of publications; and improve and support training, as well as to search for ways to increase foreign collaboration.

Finally, we must not lose the commitment and dynamic of work among people who are conscious of the importance of this paradigmatic shift in agriculture, especially with the threat that the new paradigm will face when imported chemical inputs become widely available again, given that there still are many farmers who use organic methods out of necessity rather than conviction.

References

ACTAF. 2000. *Desafíos de la Agricultura Orgánica para los países en desarrollo. La experiencia cubana al alcance de todos.* Asociación Cubana de Técnicos Agrícolas y Forestales.

Bergard, L.W., F. Iglesias, and M. Barcia. 1995. *The Cuban Slave Market 1790–1880.* Cambridge, MA: Cambridge Latin American Studies.

Carson, R. 1962. *Silent Spring.* Boston: Houghton Mifflin.

Castro, F. 1975. *La Historia Me Absolverá.* Havana: Editorial de Ciencias Sociales.

Colborn, T., D. Dumanoski, and J.P. Myers. 1996. *Our Stolen Future: Are We Threatening Our Fertility, Intelligence, and Survival?: A Scientific Detective Story.* New York: Dutton.

Companioni, N., A. Rodríguez Nodals, M. Carrión, R.M. Alonso, Y. Ojeda, and E. Peña. 1997. *La agricultura urbana en Cuba: Su participación en la seguridad alimentaria.* Conferencias. III Encuentro Nacional de Agricultura Orgánica. Universidad Central de Las Villas. Villa Clara, Cuba: 9–13.

Durán, J.L. 1998. Degradación y manejo ecológico de los suelos tropicales, con énfasis en los de Cuba. *Agricultura Orgánica* 4:1:7–11.

Ellwood, W. 1998. Cuba: The facts. *The New Internationalist* 301:24–25.

ENCC. 2000. *Datos de Archivo. Estación Nacional de Café y Cocoa.* Havana: MINAG.

Espinosa, E. 1997. La economía Cubana en los 1990: De la crisis a la recuperación. *Carta Cuba.* Facultad Latino Americana de Ciencias Sociales. Havana: University of Havana.

Francisco, I. 2000. Trabajan 96.2 de cada cien cubanos laboralmente activos. *Granma* 36:253:3.

Funes, F. 1997. Experiencias Cubanas en agroecología. *Agricultura Orgánica* 3:2–3:10–14.

Funes, F., M. Monzote, and F. Funes-Monzote. 1999. Perspectivas de la Agricultura Orgánica en Cuba. Documento presentado al Consejo Técnico Asesor del MINAG.

García, L. 1999. Educación y capacitación agroecológica. *Agricultura Orgánica* 5:3:9–12.

García, L., N. Pérez, and E. Freire. 1999. Centro de Estudios de Agricultura Sostenible: Su contribución a la difusión de la agricultura orgánica en Cuba. *Agricultura Orgánica* 5:3:13–16.

Hernández, A., R. Ramos, and J. Sánchez. 1998. La yuca en asociación con otros cultivos. *Agricultura Orgánica* 4:2:20–21.

IICF. 2000. Datos de Archivo. Instituto de Investigaciones de Cítricos y Frutales, MINAG.

IM. 2000. Datos de archivo. Instituto de Meteorología, CITMA.

INISAV. 2000. Datos de Archivo. Instituto de Investigaciones de Sanidad Vegetal, MINAG.

Lane, P. 1997. *El modelo cubano de desarrollo sostenible.* Seminario Internacional Medio Ambiente y Sociedad. Havana.

Lappé, F. M., J. Collins, and P.M. Rosset, with L. Esparza. 1998. *World Hunger: Twelve Myths.* Second edition. New York: Grove Atlantic.

MINAG. 1999. *Datos básicos.* Ministerio de la Agricultura. Cuba.

Monzote, M. 1999. Iniciativas para la educación ambiental no formal. *Agricultura Orgánica* 5:3:19–21.

Monzote, M. and F. Funes. 1997. *Agricultura y Educación Ambiental.* Primera Convención Internacional sobre Medio Ambiente y Desarrollo. Memorias Congreso de Educación Ambiental para el Desarrollo Sostenible. Havana.

Monzote, M. and F. Funes-Monzote. 1997. Integración ganadería-agricultura. Una necesidad presente y futura. *Agricultura Orgánica* 3:1:7–10.

Moreno-Fraginals, M. 1978. *El Ingenio.* Havana: Editorial de Ciencias Sociales.

Muñoz, E. 1997. Principios y fundamentos de la integración agrícola-ganadera. *Agricultura Orgánica* 3:1:7–10.

Nova, A. 1997. *Hacia una agricultura sustentable.* Conferencias. III Encuentro Nacional de Agricultura Orgánica. Universidad Central de Las Villas. Villa Clara, Cuba: 4–8.

Peláez, O. 2000. Trabajan por acelerar programas de reforestación. *Granma* 36:258:2.

Pérez, N. 1997. Bioplaguicidas y Agricultura Orgánica. *Agricultura Orgánica* 3:2–3:19–21.

Riverol, M. 1998. Bordos de desagüe, una tecnología para reducir las pérdidas de suelo. *Agricultura Orgánica* 4:1:18.

SANE. 1999. *Informe final Proyecto SANE (Sustainable Agriculture Networking and Extension).* ACAO/CLADES/PNUD.

Sánchez, L. and L. Chirino. 1999. "De campesino a campesino": Apuntes para una propuesta. *Agricultura Orgánica* 5:3:24–27.

Serrano, D. 1998. Uso de policultivos en sistemas integrados agricultura-ganadería. *Agricultura Orgánica* 4:2:22–23.

Socorro, M., L. Alemán, F. Cruz, J. Deus, R. Cabello, and A. García. 1997. *El cultivo del arroz en Cuba en el contexto de la agricultura orgánica.* Conferencias. III Encuentro Nacional de Agricultura Orgánica. Universidad Central de Las Villas. Villa Clara, Cuba: 82–84.

Varela-Pérez. 2000. Fabricación de más de 20 tipos de azúcares en la actual zafra. *Granma* 36:280:2.

CHAPTER TWO

Cuban Agriculture Before 1990

Armando Nova, Center for Research on the Cuban Economy, University of Havana

Before 1959 Cuban agriculture was characterized by the ubiquitous presence of foreign capital, its fusion of self-interest with local agricultural and financial oligarchies, and by an extreme concentration of landholdings in large sugarcane plantations and cattle ranches. Thirteen American sugar companies owned 117 million hectares of land, with an estimated 25 percent of total arable land under foreign control. Of the rest, just nine large Cuban sugarcane plantations covered more than 620,000 hectares, which together with the agricultural bourgeoisie controlled more than 21 percent of the land (1.8 million hectares). The rural middle class, lower middle class, and *campesinos* who owned their own land, had approximately 2.5 million hectares. Overall, 9.4 percent of landowners had 73.3 percent of the land, a very inequitable distribution of the means of production (see Table 1).

Table 1. Landholdings in 1959

	UNITS	TOTAL	UP TO 65 HECTARES	UP TO 400 HECTARES	OVER 400 HECTARES
Area	Hectares (percent)	8,522,276 100	628,673 7.4	1,641,440 19.3	6,252,163 73.3
Farms	No. (percent)	42,089 100	28,375 68.3	9,752 23.2	3,602 8.5
Owners	No. (percent)	30,587 100	20,229 66.1	7,485 24.5	2,873 9.4

(Acosta 1972)

Cuba had a distorted agricultural economy, essentially based on one crop and one export, as can be seen in Table 2. Of course sugar was the one crop and the one export, accounting for more than 75 percent of the total value of Cuban exports.

One consequence of this economic structure was poor living conditions, particularly among rural people. The maximum annual income of agricultural workers was less than 300 Cuban pesos, with subhuman living standards—60 percent were living in palm huts with dirt floors. There were no sanitary installations, not even simple latrines or running water. Seventy-nine percent used kerosene for light, while the rest had no nighttime illumination at all. In terms of food, only 11 percent consumed milk, 4 percent meat and 20 percent eggs, while the main staples of their diet were rice, beans, roots, and tubers. Forty-three percent were illiterate and 44 percent never attended school.

Table 2. Cuban exports 1953 to 1957 (%)

CATEGORY OF EXPORTS	1953	1954	1955	1956	1957
Durable goods	5.0	0.6	0.6	–	0.6
Non-durable goods	86.6	84.7	84.7	86.2	87.6
Fresh foodstuffs	0.8	1.5	1.9	4.5	2.7
Processed foodstuffs	78.6	74.9	74.3	74.3	78.2
Preserved foodstuffs	0.4	0.3	0.4	0.5	0.7
Beverages	0.2	0.2	0.3	0.2	0.2
Tobacco	6.5	7.6	7.3	6.6	5.9
Other	0.1	0.2	–	–	0.1
Fixed capital goods	0.2	0.2	0.3	0.3	0.4
Intermediate capital goods	12.7	14.5	14.9	12.8	11.4

(DGE 1957)

In the pre-Revolutionary period, agriculture was a mixture of semi-feudal remnants combined with capitalist practices. The remnants of feudalism included payment with coupons for the company store, and the use of the army for labor control. Among the capitalist features were salaries, new methods of organization, and use of modern implements and technology. While the prices that small farmers received for the crops were low, intermediaries and middlemen made large profits at their expense (see Table 3).

Table 3. Prices of agricultural products in Cuba (price in Cuban pesos)

PRODUCT	FARM	WHOLESALE	RETAIL	RETAIL/FARM RATIO
Peppers	2.8	15.0	20.0	7.2
Squash	0.6	3.5	5.0	8.1
Tomato	1.4	30.0	45.0	32.6
Pineapple	3.1	20.8	30.0	9.8
Avocado	1.0	9.0	15.0	14.8
Oranges	0.9	2.7	4.0	4.4

(JNE 1953)

The long historical process of distorted development had by the 1950s converted Cuba into a supplier of raw materials, mainly sugar, and a buyer of all kinds of goods, especially from the United States—even though the conditions might be favorable for producing these goods within the country (see Table 4). Above all this produced a great dependence on the American market.

Table 4. Imports as a percentage of total consumption

ITEM	PERCENT
Edible fats	88
Vegetables	33
Cereals	40
Meat products	63
Canned fruits	84

(Adapted from different sources)

Cuba After 1959

Under the first and second Agrarian Reform Laws the Cuban state took control of more than 70 percent of the arable land and created the state sector in agriculture. The area nationalized reached 5.5 million hectares of which 1.1 million were turned over to those working the land, leaving the state in control of approximately 71 percent of the total area.

The existence of the large state sector made a planned reorganization of land use possible (Vilariño and Domenech 1986). The strategy was always to diversify agriculture to reduce the dependence on one product, sugar, increasing the variety of foodstuffs exported, and to substitute national production for imports. When the United States cancelled Cuba's sugar quota—one of the first actions taken against the Cuban revolution—and given the diversification policy, it was decided to reduce the area devoted to sugarcane. Nevertheless, the ex-Soviet Union and the rest of the socialist countries in Eastern Europe decided to purchase Cuban sugar in bulk, thus creating a secure market, with long-term, stable, and preferential prices. This led to a decision to reconsider the reduction of area devoted to sugarcane, thus prolonging our dependence on a one-product farming system.

A number of other factors contributed to the revised direction of policy, including:

- the ideal natural conditions for sugarcane production
- the vast knowledge and experience in sugarcane and sugar production
- the huge installed industrial capacity and investment already made in sugar processing

The policy directions followed in the first years of the Revolution regarding the use of nationalized land were clearly expressed by Fidel Castro at the closing session of the First Farm Workers Congress in February 1959: "To maintain consumption, to maintain abundance, to carry out agrarian reforms, the land cannot be distributed in one million small pieces. . . cooperatives must be established in the right places for each type of production, and the crops to be sown must be planned. . ." (Castro 1959).

The National Institute for Agrarian Reform (INRA) was created to be in charge of the application and enforcement of the Agrarian Reform Law. Given the characteristics of the nationalized plantations and ranches, two systems were created to organize production: so-called "people's farms" on former cattle ranches and virgin lands, and cooperatives on the sugarcane plantations. After the 1960 harvest a large portion of the expropriated sugarcane areas were transformed into sugarcane cooperatives, where the state still owned the land and other means of production, while the workers tilled the land in usufruct. In late 1962, policy makers decided that these cooperatives were not working out, and they were transformed into "People's State Farms." When the Agricultural Enterprises were created in 1963 to organize state production, there were approximately 272 People's State Farms, 613 sugarcane cooperatives, and 669 Administrative Farms (formed directly from expropriated plantations). By the end of 1964, 263 new enterprises had been established.

During the period from the First Agrarian Reform Law until 1975, no important changes occurred in the collective organization of production among small landholders, except for the creation of Credit and Service Cooperatives (CCSs) and agricultural communities. Then at the Fifth Congress of the National Association of Small Farmers (ANAP), following up on decisions made at the First Congress of the Cuban Communist Party, the collectively run Agricultural Production Cooperatives (CPAs) were created. By 1998 some 1,139 cooperatives of this type had been formed, covering some 710,000 hectares (an average of approximately 625 hectares per cooperative) with more than 63,000 members. Development of the CCSs continued as well; they now cover some 980,000 hectares with more than 168,000 members. Finally, there are approximately 250 farmers' associations (more loosely organized than CPAs or CCSs) with more than 9,400 members, covering an area of more than 26,000 hectares.

In transferring most of the expropriated lands to the state, the aim was to accelerate the adoption of advanced technologies and boost productivity, while maintaining a non-exploitative labor system. This would be the starting point for the establishment of large agricultural enterprises. The more just distribution of wealth, and the new relations of production which were favorable to the development of the productive forces, led to the sustained growth of agricultural production over many years.

Sectoral Analysis: Food Production

In the food sector, the first priority was to boost cattle, milk, and egg production in search of fast and economical ways to immediately supply animal protein to the population.

CATTLE AND DAIRY PRODUCTION

To increase milk production, approximately 20,000 dairy cows as well as high-quality bulls were imported to improve dairy herds. Investments were made in technology with the establishment in 1961 of artificial insemination trade schools, and the large extensions of grazing land were subdivided, improved types of pasture grasses were sown, and hay and silage were produced.

Between 1961 and 1965 important transformations in cattle ownership took place, as shown in Table 5. The transition toward greater state ownership took place as private farmers reared and improved beef cattle, which were later sold to the state. At the same milk production was being built up in the state sector.

Table 5. Head of cattle by sector, 1961 to 1965

	AUGUST 1961		DECEMBER 1965	
	TOTAL	PERCENT	TOTAL	PERCENT
National total	5,776	100	6,700	100
State owned	1,400	24	3,844	57
Privately owned	4,376	76	2,856	43

(González and Miranda 1984)

The technical and material basis for milk production was consolidated between 1962 and 1970. In 1970 a genetic transformation process of the cattle stock was initiated. At that time approximately 50 percent of the females had marked dairy traits, and from there on over the next several years 900,000 cows were inseminated each year. By 1976 the state sector had a predominantly dairy stock, a situation that is still true today. This boosting of dairy characteristics led to an important growth in milk production (Table 6). However, while the stock was improved genetically, it now required more feed in order to exploit its higher productive capacity.

Table 6. Production of fresh milk per year

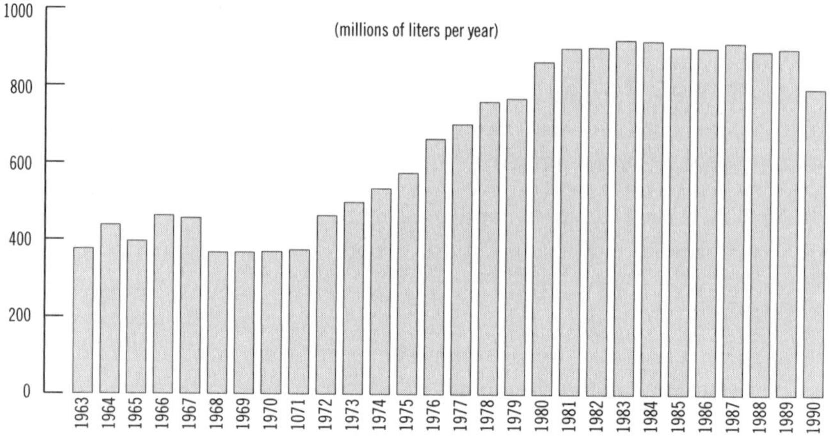

(millions of liters per year)

1000
800
600
400
200
0

1963 1964 1965 1966 1967 1968 1969 1970 1071 1972 1973 1974 1975 1976 1977 1978 1979 1980 1981 1982 1983 1984 1985 1986 1987 1988 1989 1990

Note: Data from 1968 on refer only to the state sector. The private sector produces an annual average of 150 million liters.

(González and Miranda 1984; MINAG 1983, 1991)

As the average income levels among the Cuban population began to grow, so did the demand for foodstuffs. This in turn led to a decision in 1962 to institute food rationing to guarantee equal access. As rationing included the meat sold to the population, strict control had to be instituted over slaughtering in order to protect the cattle stock. As a consequence of this policy, slaughtering numbers decreased. This together with technological improvements allowed the national cattle herd to reach 7.2 million head in 1967, the highest number in the history of Cuban livestock production.

In spite of the 1962 steps taken to preserve the stock, additional measures were implemented in 1973 to stem a several-year decline to 5.5 million head. The steps taken included tightening the beef supply to the population, cutting by 50 percent the ration for students with scholarships, since they received free extra food at school, and substituting beef with poultry and pork. These measures helped the cattle herd to surpass 5.6 million head in 1975, at which point another long-term decline set in, which has continued to the present day.

Fresh milk production recorded significant increases from 1963 to 1983. Throughout the 1980s however, production stagnated, with a decreasing trend after 1984, due to a reduction in the number of milking cows, though the number of liters per cow remained constant. For beef cattle, the average weight of beef produced per hectare declined from about 338 kg in 1981 to 325 kg in 1988. The extraction rate of the herd averaged just 18 percent from 1980 to 1988, typical for extensive livestock production, when the indicator for intensive livestock production should range between 25–40 percent. The problems in this sector have generally been attributed to low pasture quality—from which cattle typically obtain about 70 percent of their nutritional requirements—in part because the best lands were transferred to crops like sugarcane, leaving pastures on low quality soils, where they received insufficient fertilization and irrigation.

SWINE PRODUCTION

Before 1959 the rearing of pigs was generally concentrated in the hands of small farmers, though on some larger ranches swine were produced using harvest residues as feed. In 1960–1961 attempts were made to develop pig production. Thirty thousand sows were imported from Canada to guarantee the genetic basis, but it was not until 1968 that production began to rise as did the number of animals.

In 1971 an outbreak of African swine fever in Havana province led to the slaughter of more than 80,000 head in the state sector, and more than 230,000 in the private sector. A series of sanitary measures were implemented to eradicate the disease, and after that, private pig rearing was prohibited in Havana province until 1990. In 1978, a second focal point of disease was detected in the western province of Guantánamo, and a series of measures similar to those adopted in 1968 were taken to completely wipe out the swine fever.

The number of swine rose for several years, as did the production of pork, which reached 102.4 million tons. However, from 1986 on the efficiency of production declined (see Table 7).

Table 7. Efficiency indicators of swine production, 1985 to 1989

YEAR	DEATHS PER 1000 HEAD	PIGLET MORTALITY PERCENT	AVERAGE WEIGHT, KG	FEED CONSUMED (METRIC TONS)		
				SOLID FEED	MOLASSES	LIQUID FEED
1985	100.5	7.1	89.0	295.0	193.3	874.0
1986	103.7	7.6	86.3	313.2	151.5	952.3
1987	119.5	8.3	85.8	341.9	181.6	1,015.1
1988	132.3	8.9	86.3	361.1	209.1	1,082.8
1989	160.0	9.4	85.1	397.5	226.5	1,200.0

(CEE 1989)

POULTRY PRODUCTION

In 1963 a program was initiated to boost poultry production. Eggs from purebred chicken lines were imported for breeding, to form the necessary genetic base for production. The National Poultry Complex (CAN) was created in 1964 as a vertically integrated enterprise. Poultry rearing, a priority sector within agriculture, sustained production increases in subsequent years, with egg production at the forefront (see Table 8). The efficiency of this industrial style of production peaked in 1986, after which there was a decline (Nova and González 1990).

Table 8. Dynamics of poultry production, 1962 to 1989

YEAR	NUMBERS		LAYING INDEX (eggs/head)	EGG PROD., MILLIONS	POULTRY MEAT, METRIC TONS	FEED CONVERSION RATE		VIABILITY OF CHICKS, PERCENT
	CHICKS, 1000s	LAYERS, 1000s				KG/10 EGGS	KG/KG MEAT	
1962	5,222	640	–	179	–	–	–	–
1965	–	–	209	915	–	1.86	–	–
1967	2,505	5,707	196	1,667	24	1.94	–	94.4
1970	1,121	7,066	197	1,426	20	1.94	3.4	91.3
1975	4,488	7,028	235	1,851	56	1.66	2.8	93.7
1980	7,819	8,658	246	2,326	90	1.60	2.9	94.0
1985	9,469	8,476	249	2,523	113	1.54	2.5	94.8
1986	9,541	9,033	246	2,519	113	1.55	2.5	94.2
1987	8,945	9,088	245	2,495	109	1.62	2.6	93.2
1988	9,700	9,005	236	2,460	114	1.69	2.6	93.0
1989	–	9,917	222	2,522	117	1.62	2.7	91.9

(CEE 1974, 1980, 1985, 1989)

ANIMAL FEED

The production of animal feed rose from an annual average of approximately 1.5 million metric tons in 1981–1985 to about 1.9 million metric tons in 1986–1990. Feed production over this entire period was characterized by a heavy dependence on imported ingredients, as shown in Table 9.

Table 9. Proportion of imported ingredients in animal feed

	PROPORTION IMPORTED, PERCENT			
	1981–1985		1986–1990	
	IMPORTED	NATIONAL	IMPORTED	NATIONAL
Basic feeds	85	15	85	15
Plant protein	71	29	86	14
Animal protein	58	42	60	40
Phosphorated feeds	88	12	93	7

(García and Fernández 1990)

RICE PRODUCTION

Rice production has a long tradition in Cuban agriculture, and plays an important role in the daily diet of the Cuban consumer. References indicate that in 1862 Cuba produced about 50 percent of its national rice requirements. The rise of the large sugarcane plantations and cattle ranches, and the interests of American producers and exporters, affected the development of rice production. During and after World War II, difficulties in shipping and supplying rice led to reduced imports and stimulated domestic production.

After 1951 planting areas were increased to approximately 54,000 hectares, jumping to 113,707 hectares in 1958, producing 135,000 metric tons of rice, with an average yield of 1.86 tons of dry hulled rice per hectare. During this period some 190,000 tons/year were imported, yielding a per capita consumption of approximately 50 kg/year. Though rice production was highly mechanized and had acceptable levels of water and fertilizer inputs, yields reached only 79 percent of that achieved by American rice growers.

After the triumph of the Revolution, a program was put in place to increase production. But shortly thereafter, in 1964, reliable imports began from the People's Republic of China and rice areas and production levels were reduced. In 1968, when relations between Cuba and China deteriorated, the rice program was restarted. Large investments were made in canals, irrigation systems, and industrial processing facilities. High-quality seed lines were imported, as were chemical fertilizers and pesticides.

Until the 1990s an average of 142,000 hectares were harvested yearly, producing about 400,000 metric tons of dry hull rice, for a yield of about three metric tons per hectare (an increase of nearly 60 percent over 1958), and production covered approximately 50 percent of national needs.

OTHER FOOD CROPS

Roots and tubers, beans, corn, plantains, tropical fruits, and vegetables are all-important crops providing a significant part of the staple diet. According to the 1946 agricultural census, bean production then covered two-thirds of national needs. The demand for black beans was completely covered by national production, while red bean production covered 50 percent, and other varieties, such as white beans, chickpeas, and peas were imported from Chile, the US, and Mexico.

Beans traditionally grown in Cuba reached production levels of 59,500 and 55,700 tons between 1961 and 1962, respectively. Then a decreasing trend began, falling to between 8,000 to 14,000 tons/year, which led to an annual importation of more than 100,000 tons, at an annual expense of more than 40 million pesos (García and Nova 1987).

Corn production levels were also historically significant and they remained so after 1959. Corn production is primarily for fresh consumption. Annual imports of dry corn for animal feed rose to some 400,000 tons, at an estimated price of 40 million pesos.

In 1946 very low levels of vegetable production were recorded, perhaps reflecting a low dietary preference for them. Today supplies are much higher, and vegetables make a significant contribution of vitamins and minerals to the diet.

Sectoral Analysis: Export Crops

THE SUGAR INDUSTRY

The sugar industry has traditionally been highly productive in Cuba. During the colonial period the Spanish found favorable natural conditions and introduced sugarcane. By the end of the seventeenth century sugar was already being processed in Cuba.

Sugarcane production passed through a number of stages between the turn of the century and 1958 (see Table 10). The highest levels of production in the pre-Revolutionary stage were registered in the 1950s.

Table 10. Periodization of sugarcane production from 1902 to 1958

STAGE	PERIOD	YEAR	1000s METRIC TONS	STAGE	PERIOD	YEAR	1000s METRIC TONS
Beginning of the republic	1902–1911	1902 1906	0.850 1,178	pre-WW II	1936–1939	1936 1939	2,557 2,723
pre-WW I	1912–1914	1912 1914	1,845 2,598	WW II	1940–1945	1940 1945	2,779 3,515
WW I	1915–1919	1915 1919	2,609 4,009	post-WW II	1946–1950	1946 1950	4,011 5,492
post-WW I	1920–1927	1920 1924 1927	3,735 4,113 4,508	1950s	1951–1958	1951 1952 1958	5,690 7,138 5,862
Critical Period	1928–1930	1928 1935	4,041 2,538				

(AAC 1958)

Before the Revolution sugarcane covered approximately 2.68 million hectares, of which some 75 percent belonged directly to sugarcane mills and the rest were leased. 1.2 million hectares of cane were cut each year. About 360,000 hectares were held in reserve each year, depending upon the price of sugar in the world market.

In light of the new market consisting of the Soviet Union and other socialist countries, sugarcane production was revitalized a few years into the Revolution. Large investments were made, especially at the end of the 1960s (Nova 1988a), to try to reach the goal of 10 million metric tons in 1970, which was not attained, though it was the all-time record year with 8.5 million tons.

In the pre-Revolutionary period, yields per area of sugarcane had been low. In 1965 yields began a positive trend, eventually rising by more than 37 percent. However, they still lagged in comparison to yields obtained by other world producers. In 1988, the yield in state-owned areas was of 55.9 tons per hectare, while in the private sector it was 61.3 per hectare (CEE 1988).

An important achievement was the mechanized cutting and harvesting of sugarcane. At present, more than 66 percent of the area is mechanically cut, and 100 percent is mechanically harvested. This helped boost productivity, as well as humanizing the work in the field (Cruz et al. 1989). In 1979 a rust disease, *Puccinia melanocephala*, was detected infecting cane in Cuba. In 1980 the affected areas were replaced and replanted with resistant sugarcane varieties.

The comparative advantages of producing sugar—favorable natural conditions and secure markets with no deterioration in the terms of trade—contributed greatly to the continued dependence on one product in the world market. To reduce the dependence of sugar exports, for years work was targeted at the development of industrial sugarcane by-products (Noa 1977). Nevertheless, in 1988 sugar itself was still the main product being exported, accounting for 74 percent of exports (CEE 1988).

CITRUS INDUSTRY

Citrus fruit production began during Spanish colonization. During the republic it remained poorly developed and dispersed for a long time, as there were no internal stimuli and the powerful American citrus industry controlled the world market. In 1958 citrus orchards covered approximately 12,000 hectares and production reached about 60,000 tons. After 1959 domestic production grew with the establishment of new plantations, the recovery of existing orchards, and a constant improvement of agronomic practices (Nova 1988b).

The first stage of citrus fruit development in Cuba culminated during the first years of the Revolution, peaking in 1968. However, there still was no defined citrus development program; citrus production was a part of a general plan for the diversification of agriculture. In 1968 an Integrated Development Program for citrus was put into operation. By 1989 145,000 hectares had been added, representing 63 percent of the total area sown during the period from 1959 to 1989. In 1990 more than one million tons of citrus were harvested, 16 times the level reached before 1959, and made up three percent of the total value of Cuban exports.

Among the main objectives of the citrus program was also a contribution to the improvement of the Cuban diet; today per capita consumption of fresh fruit is approximately 2.1 times higher than in 1959. Through the citrus program new industries were constructed for processing concentrated and natural juices, essential oils, jams, and other products. By the beginning of the 1990s there were three industrial complexes with a processing capacity of 104 tons of input per hour.

The citrus fruit program required major investments, which made it possible to achieve a technological level comparable to that found internationally. However, although agricultural yields per hectare are double those obtained before 1959, they are still low (about nine metric tons per hectare at the beginning of the 1990s).

The citrus program was initially developed in the context of the Council for Mutual Economic Assistance (COMECON), the economic union of the former Socialist Bloc. The radical change in trade relations with the disappearance of COMECON brought about a reorientation of citrus production, increasing the importance of processing within the citrus production system. The development of the Cuban citrus industry must be based on cost effectiveness for it to be competitive, and to be able to reengage in the international market (Nova and González 1990).

Conclusion

Between 1959 and the beginning of the economic crisis or Special Period there were notable increases in Cuban food production. However, at the end of the 1980s there was a generalized decrease in yields and in other indicators of efficiency in an important group of commodities. This came about in an intensive development model, based on high levels of external inputs and a high external dependence (mainly machinery, fuel, and agrochemicals); similar to the situation faced by other countries applying the same productionist model (Rosset 1997). Furthermore, the quantities produced were not sufficient to fully cover the demands of the population with any economic effectiveness. Meanwhile a very significant proportion of arable land was used for export production (see Table 11), and many soils had begun to show signs of degradation (salinity, erosion, acidity, poor drainage, etc.). These factors already made it important to carry out economic, structural, technical, and organizational transformations in Cuban agriculture. The events that occurred in the Eastern European countries only made the task more urgent.

Table 11. Arable land use, 1989

USE	% OF TOTAL	% STATE	% PRIVATE
Exports	53	54	48
Foodstuffs	44	43	48
Other	3	3	4
Total	100	100	100

(Adapted from CEE 1989)

References

Acosta, J. 1972. Las leyes de reforma agraria en Cuba y el sector privado campesino. *Revista Economía y Desarrollo* 12.

AAC. 1958. *Anuario Azucarero de Cuba (Varios hasta 1958).* Havana: Comité de Estabilización del Azúcar.

Castro, F. 1959. Discurso pronunciado en clausura del I Congreso Campesino. Havana.

CEE. 1974, 1980, 1985, 1986, 1987, 1988, and 1989. *Anuario Estadístico de Cuba.* Havana: Comité Estatal de Estadísticas.

Cruz, V., N. Torres, and S. Aguerreberre. 1989. Análisis de los diversos aspectos de la mecanización agrícola. *Revista Cuba Economía Planificada* 3.

DGE. 1957. *Agricultura. Anuario Estadístico de Cuba.* Dirección General de Estadísticas. Havana: Ministerio de Hacienda.

García, A. and P. Fernández. 1990. *La ganadería cubana: análisis del decenio 1981–1990 y algunas consideraciones sobre su desarrollo perspectivo.*

García, C. and A. Nova. 1987. Importancia y perspectiva económica de la producción del frijol en Cuba. *Revista Economía Planificada.*

González, C. and R. Miranda. 1984. *Economía Agropecuaria.* Havana: Editorial Pueblo y Educación.

JNE. 1953. *Junta Nacional de Economía.* Havana: Instituto Nacional de Reforma Económica.

MINAG. 1983. *Series históricas* (Sector estatal) 1971–1982. Havana: Viceministerio de Ganadería.

MINAG. 1991. *Series históricas* (Sector estatal) 1980–1991. Havana: Viceministerio de Ganadería.

Noa, H. 1977. La industrialización de los derivados como apoyo a la economía de producción azucarera. *Sobre los derivados de la caña de azúcar* 11:77.

Nova, A. 1988a. Apuntes sobre el proceso inversionista en la actividad agropecuaria. *Revista Economía y Desarrollo* 4.

Nova, A. 1988b. *Aspectos económicos de los cítricos en Cuba.* Havana: Editorial Científico-Técnica.

Nova, A. and C. González. 1990. *La organización agroindustrial en Cuba.* Havana: Editorial Científico-Técnica.

Rosset, P.M. 1997. *La crisis mundial de la agricultura convencional y la respuesta agroecológica.* Conferencias. III Encuentro Nacional de Agricultura Orgánica. Universidad Central de Las Villas. Villa Clara, Cuba: pp. 87–95.

Vilariño, A. and J. Domenech. 1986. *El Sistema de Dirección y Planificación de la Economía Nacional, historia, actividad y perspectiva.* Havana: Editorial Pueblo y Educación.

CHAPTER THREE

Cuban Agriculture and Food Security

Marcos Nieto, Negotiating Group, Ministry of Agriculture, and Ricardo Delgado, Cuban Association of Agricultural and Forestry Technicians

The Economic Context

If one were to look at the first four decades of the Cuban Revolution, what stands out are the positive steps taken by the Cuban government to assure food availability. The priority given to food security, and the centrally planned economy, aided efforts toward meeting the food requirements of the Cuban population, even though there was always some gap between expectations and results, especially as climatic conditions for crop production led to regional differences in meeting the needs of the population.

During the second half of the 1960s, the Cuban economy achieved adequate levels of development, with average annual per capita GDP growth of more than 2.5 percent. The efforts of those years established a high industrial capacity, new infrastructure (roads, airports, ports, and bridges) and placed great importance on human resources, which would be the basis and source of our future economic development.

At the end of the 1980s, the Cuban government initiated a new era of renovation that was known as "Rectification of Errors and Negative Tendencies." Key sectors were identified and obstacles to efficiency were diagnosed and analyzed in open debate, while recognizing the need for new forms of property. Before new policies could be completed and implemented, however, the collapse of the Eastern European socialist countries occurred, with drastic consequences. There was a marked reduction in commercial trade, and credits disappeared, with an enormous impact on strategic development.

Between 1989 and 1993 there was a 75 percent reduction in imports and 79 percent in exports. In import terms, this meant losses due to higher prices. Investments were paralyzed or reduced to zero, causing stagnation in many sectors of our economy, with irreparable consequences. Food imports declined sharply, along with oil, raw materials, and supplies of spare parts. The economy was severely affected; the government named this era the "Special Period."

The budget deficit in 1998 reached 559.7 million pesos, and while the currency supply in the hands of the population was way beyond the need for capital circulation (an inflationary and shortage-generating currency "overhang"), the situation was not as critical as in 1993, when average cash on hand reached 15.4 months' salary, and consumers could not find sufficient goods in stores on which to spend their salary.

Foreign exchange was also restricted, with a lack of foreign currency and an irregular financial situation leading immediately to a rapid increase in the foreign debt. Short-term financial credits with high interest rates, no credits for development projects, and the intensification of the US blockade completed the scenario.

As a result, attention focused on the following measures:

- a priority on activities producing exportable goods or the substitution of imports, and those promoting internal foreign currency sales of goods and services for tourism and the presence of foreign capital
- energy savings and use restrictions
- making agriculture and food production the first priority
- support development of the pharmaceutical and biotechnological industries

These, among other measures, formed the main strategy for the process of transformation. Studying ways in which to carry this out, and with a broad process of popular consultation, Cuba faced a profound restructuring, adapting itself to the new global conditions as it joined the international market.

The strategy led to a mild economic recovery that stopped the recession in 1994, with a slight increase in GNP of 0.7 percent, a second increase of 2.5 percent 1995, finally reaching 4.2 percent growth in 1999. Seven years after deep reforms were initiated, after analyzing the measures taken, one can have some idea of the present Cuban economy. The restructuring of foreign trade is reflected in a strong reorientation toward Latin America and Canada, followed by Europe and Asia. Also evident is the decentralization of foreign trade, with a tripling of the number of marketing entities. Exports began to recover in 1994 (15.6 percent growth), and again in 1995 (12.5 percent), while in 1998 exports reached 15.8 percent of GNP.

Foreign investments have played an important role by creating new markets and increasing product competitiveness through technology and capital. While foreign capital and the decentralizing of foreign trade are two elements in the new financial situation, they are not the only ones. The presence of foreign capital, considered an economic agent since 1983, was reinforced by constitutional reforms in 1992, and is legally established in economic policy with new procedures and guarantees made explicit in 1995.

Structural changes annulled the state monopoly over foreign trade, thus granting recognition to the Cuban companies and joint venture corporations that are part of the new economic system. This is being accompanied by restructuring the organization of the state toward decentralization, eliminating bureaucracy, and granting wider flexibility.

The contraction of the central state apparatus that began in April 1994 (a reduction from 50 to 32 entities), has strong effects on functional simplification, the improvement of planning and financial control, and the implementation of new administrative techniques.

The state restructuring process has now reached basic levels in two ways. First, industrial redimensioning has been directed towards the reorganization of production flows, the adjustment of production capacities, and a concentration on efficiency. Second, restructuring aims to improve business practices in order to meet contractual obligations, and achieve functional autonomy and financial self-sufficiency at the level of the firm.

Different methods to stimulate the circulation and acquisition of foreign exchange have included the development of joint-venture corporations, the authorization of direct importation, the growth of tourism, the emergence of new financial schemes, the increase in family remittances, and incentives paid in convertible currency (affecting more than one million workers). The use of foreign exchange in Cuba has been authorized for these purposes, and the possession of foreign currency has been decriminalized.

The restructuring of agricultural landholdings has been perhaps the most transcendent organizational change. In a brief time a large number of new cooperative farmers and *campesinos* were added to the previously scarce number of cooperatives, collectives, and private farmers. Forty-two percent of the land belonging to the state was turned over to new cooperatives and private farmers. Today, 67.3 percent of agricultural land is in the hands of private farmers, representing 40.7 percent of the total area of Cuba.

Large farms and state enterprises have been redistributed to those interested in cultivating the land in free usufruct, in almost 3000 Basic Units of Cooperative Production (UBPC). At the same time new production incentives are being offered, and traditional agriculture is encouraged in crops for which the conventional agricultural model no longer works due to the lack of imported resources. Land has been given to thousands of families for the

A farmers' market in Havana.

production of coffee, cocoa, tobacco, and other crops. There has been a net resurgence of population in the rural sector, enabling growth in agricultural and livestock activities.

All producers have purchase contracts with the state, guaranteeing the distribution of food staples to the entire population. In addition, farmers' markets were created in 1994, where the farmers can sell their production surpluses at prices set by supply and demand. The new markets not only motivate private and cooperative farmers, but also workers on remaining state farms, since they can receive higher incomes and participate in the profits.

With the recovery of production and services, and the opening of agricultural and industrial marketplaces, excess money supply could be taken out of circulation and the Cuban peso was revalued. At the beginning of 1994, the rate for a dollar on the parallel market was 150 pesos, but by the end of 1999, it had fallen to 20 pesos to the dollar.

As a result of these measures and the constant growth in production in recent years, the national budget deficit was reduced from a high of 1,500 billion pesos in 1993 to 14.2 billion pesos 1994, 765.5 million in 1995, 580 million in 1996, and 268 million pesos in 1998. Today the deficit stands at approximately two percent of the GNP, down from 33 percent in 1993.

Agriculture and Food

Of a total of 11 million hectares, 6.7 million hectares are suitable for agriculture; by the end of 1998 4.5 million hectares were cultivated. Of the land not suitable for cultivation, forests cover 2.6 million hectares and the rest is used for reservoirs (0.5 million hectares, representing more than 90 percent of the water reserves in Cuba).

The water supply for agriculture has been considerably improved by the construction of reservoirs and micro-reservoirs that, together with the adequate management and exploitation of underground water, allows for the use of 5.5 billion cubic meters of water per year for crop production (except sugarcane), covering an area of 564 million hectares.

Despite progress in Cuban agriculture between the 1960s and 1980s, food sufficiency and diversity to meet the population's basic needs was never achieved. Thus at the end of the 1980s, Cuba imported approximately 50 percent of the food needed to cover the basic requirements of humans and livestock. Because the production levels reached during the era of economic stability still did not satisfy basic needs, the government launched an ambitious National Food Program (PAN), which was seriously interrupted at the start of the Special Period.

The Impact of the Crisis

The economic crisis at the beginning of the 1990s led to a reduction in agricultural production, since it had been based on imported inputs such as fuel, fertilizers, pesticides, high-tech equipment, etc. Nevertheless, under these severe conditions, the production of some crops actually increased. According to official statistics, various crops and livestock production experienced a reduction in output, especially those items high in protein such as milk, beef, and eggs (see Table 1). The possibilities of short-term recovery in this area depend on the improvement of the financial situation of the country. However, significant increases in roots, tubers, and vegetables—key elements in the Cuban diet—have recently been obtained.

Table 1. Principle commodities, production in 1989 versus 1998, and percent change

PRODUCT	UNITS	1989	1998	1998/1989, %
Roots and tubers	1000 tons	972.6	1107.1	114
Vegetables	1000 tons	610.2	702.9	116
Beans	1000 tons	14.1	25.5	181
Rice (wet husks)	1000 tons	536.4	337.9	63
Citrus, fresh	1000 tons	825.7	655.6	79
Other fruits	1000 tons	218.9	112.3	51
Milk	Million liters	1131.3	697.7	62
Eggs	Millions	2672.6	1621.1	61
Beef	1000 tons	289.1	137.3	48
Poultry	1000 tons	142.8	37.4	26

Since earlier times agricultural statistics have not included the production coming from gardens, patios, backyards, and other small holdings, because they were thought to be insignificant. Nevertheless, under the present conditions it will be essential to do so. In 1999 the urban sector alone produced more than 800,000 tons of farm products, mainly vegetables. The noninclusion of this production leads to an underestimation of the production and availability of foodstuffs, which has become particularly important since 1994.

Because of the crisis, the availability of raw materials for poultry feed has been reduced by 30 percent, and at present supplies are scarce and irregular. Often farmers have had to use minimum maintenance levels of rations just to keep the animals alive while awaiting more stable solutions. Eggs, once a principal source of protein, abundant and essential to the population, were cut back by 23 percent from 1991 to 1995, with a 55 percent reduction between the best and worst years (493.1 million fewer eggs). Chicken production fell from 142,800 metric tons in 1989 to 37,800 tons in 1998. Many chickens were slaughtered because their production depended on imported feed, and it was virtually impossible to add body weight due to variable feed supplies.

There were also severe problems with milk production. In 1989 production reached 1.13 billion liters (state and private sectors) and was halved because the largely Holstein-Friesian dairy herd (good producers when they receive imported feed supplements) and their crossbreeds were not well adapted to adverse conditions. In the last few years, a modest increase in milk production has been obtained, but it is still lower than that of the previous decade. Beef production, which was also heavily dependent on imported feeds and technology, dropped from 289,000 tons in 1989 to 137,300 tons in 1998, which was due in part to the use of thousands of male animals for animal traction.

The production of roots, tubers, and vegetables fell by 10 percent from 1989 to 1994. The drop was more evident in vegetables (34 percent), since their production was heavily dependent on imported fertilizers and pesticides. In 1989 and 1995, 610,200 tons and 402,300 tons were produced, respectively. Roots and tubers increased by five percent, due mostly to the potato crop, which reached 281,600 tons in 1995, nearly equaling the 1985 record of 307,300 tons. These increases continued between 1994 and 1998, reaching a 104 percent increase over 1989 production levels in 1998. In 1999 the production of roots, tubers, and vegetables grew again, reaching 2.27 million tons. This represented 143 percent and 125 percent of the combination of these crops in 1989 and 1998, respectively.

A program to use high-tech methods such as microjet irrigation to guarantee plantain production throughout the year collapsed rapidly when supplies of fertilizers, pesticides, and fuel for irrigation systems were interrupted. Between 1992 and 1995 production fell from 515,000 to 400,000 tons.

Pork production had occupied a central place in the National Food Program, based on the development over the medium-term of a national network of integrated intensive factory-style pig production centers. This program was also interrupted by the lack of imported concentrates and other feeds.

Because rice is a staple food in the Cuban diet, it has been a particularly sensitive topic among the problems examined here. To make up for production shortfalls the government has had to spend considerable quantities of scarce foreign exchange. Rice production was stable from 1986 to 1990 at an average of 507,600 tons per year of wet rice, which crashed to 147,600 tons in 1993. Rice is a short-cycle crop that had been highly dependent on imported inputs, and thus suffered more than other crops. Recently, the new style of low-input "popular rice" production, with scarce or practically no inputs, has been able to supply half of the rice needed by the country.

Bean production (*Phaseolus vulgaris*) in Cuba does not play much of a role, since beans are almost totally imported. Citrus crops have faced severe difficulties due to the reduction in input, although in 1993 some foreign financing was obtained to boost citrus production for export, and yields have grown significantly in recent years, nearing historic highs.

From 1986 to 1990 there were high growth rates of reforestation and silviculture. When in 1987 a reforestation program was launched involving state forest enterprises and the cooperative and private farm sectors, together with the general public, tree plantations were doubled. The production of lumber increased by 40 percent, reaching 106,400 cubic meters in 1990. Wooden crates for crops, which had not been made before 1986, began to be produced, reaching 2.3 billion units. The annual planting area rose from 45,000–50,000 hectares in 1985 to 90,000–100,000 hectares per year from 1986 to 1990. Other forestry activities aimed toward improvement of existing plantations and management of natural forests had modest increases or stayed at historic levels.

With the onset of the crisis, a sharp reduction of forest industry activity and pine resin production was observed, coupled with a new high rate of firewood extraction for fuel, due to the lack of petroleum by-products. In 1993 the production of lumber dropped to 36 percent below 1990 levels; the production of wooden crates also fell by 50 percent, pine resin by 24 percent, while firewood extraction almost doubled. Coal mining, although in high demand as a substitute for kerosene, dropped by 50,000 tons due to the lack of essential inputs. Some recovery of industrial production has been observed since 1994, but it is still far from the levels obtained in 1990.

FOOD IMPORTS

An acute decline in foreign trade occurred after relations collapsed with the former Socialist Bloc, and the US tightened its embargo. The forced reorganization of trade had severe impacts, such as: lower prices for exports and higher prices for food imports, due to the sudden cancellation of trade agreements on favorable terms; reduced access to the world market; higher freight charges; severe delays in importing goods; higher costs; either no credit facilities at all, or credits on unfavorable terms with high interest rates, due both to the lack of liquid assets and to the pressure of the US government on companies and suppliers based in third countries.

Despite foreign trade being the sector most affected by the crisis, a recovery has been observed since 1994, although there have still been negative effects in the balance of payments. In 1990 foodstuffs represented 16 percent of imports, while from 1993 to 1995 they reached 28 percent of the reduced total imports. Among imported food items, cereals, grains, dry milk, and cooking oil represent the majority, both in weight and cost.

NATIONAL FOOD SELF-SUFFICIENCY

Due to a dependence on agricultural exports and basic food imports, Cuba has always ranked as food deficit country. Despite decades of heavy investment in infrastructure and the application of science and technology, agricultural activities never reached food self-sufficiency, as the country continued to rely on imported foodstuffs and ingredients for their processing, as well as on imported production inputs.

The recent changes in the Cuban economy have been exceptional in two ways. First, the magnitude of adjustments has been far greater than those experienced by other Latin American countries during the 1980s, and in Cuban history can only be compared to the 1933 economic depression. Second, the extraordinary form in which restructuring was carried out has shown the Cuban government's commitment to social equity and consensus-building. The tightening of the US blockade due to the Torricelli legislation in 1992 and the Helms-Burton bill in 1995, together with other amendments imposed by the US government in recent years, exacerbated the situation. Yet a fundamental decision was made to maintain the principles of social equity established by the revolutionary government, such as employment and wage policies, health care, education, social security, and the most equitable distribution possible of available goods and foodstuffs.

Initially the crisis was so extreme that by 1993 food consumption had fallen 30 percent from 1989. The availability of food was such that a basic diet and nutritional requirements could not be maintained, despite of the fact that food and food ingredient imports reached almost one fourth of Cuba's severely reduced total imports. Nevertheless, by 1994 food consumption began to increase through the opening of farmers' markets, the rapid growth of private production and backyard growing, and the phenomenon of shopping in "dollar stores." Together with increases in normal rural production, these factors have significantly improved food availability for the population, though some persistent problems remain, with nutritional repercussions.

According to the Institute for Food, Nutrition, and Hygiene (INHA), the average recommended daily caloric intake is 2,400 calories. While consumption has grown since 1993, in global terms it still has not reached the recommended level. According to calculations based on food production and import statistics, however, it is worth noting that these are undoubtedly underestimates, since only the most conservative estimates of private sector and backyard food production, and institutional self-provisioning are taken into account, while recent changes in land tenure have made these sources of production much more significant in alleviating food availability problems for a large part of the population.

ACCESS TO FOOD

In Cuba today the way the economy functions and the socioeconomic situation are very different in comparison to the systems and mechanisms that functioned prior to 1989. Perhaps the largest differences are found in the diversification of channels of food distribution, and the greater variety of income sources for the population.

Today, family incomes do not come solely from the relationship of the population with the state via salaried work in state enterprises or from social security programs. The development of non-state work has increased transactions between members of the population, and incomes derived from self-employment are on the rise. Also, dollar incomes have surged, due to remittances from family members outside of Cuba, special incentive systems at workplaces, and the growth of tourism and private services offered for tourists.

Incentive systems, the reduction of excess liquid assets held by the population, and the dynamics of filling the basic market basket, coupled with an increase in the supply of products for sale, have combined to stimulate people's desire to work, measured by people returning to their workplaces or seeking employment.

To help people cope with persistent food availability problems, significant food distribution channels have been kept open, such as products made available on the ration card, intentionally low meal prices at schools and workplace cafeterias, and free meals at hospitals. For part of the population, a free or low-price source of food is available via self-provisioning through home gardens, backyards, and cooperatives, and from the sale to employees at very low prices of excess production from the self-provisioning parcels maintained at workplaces to supply workers' lunchrooms.

Other sources of food—farmers' markets, fish shops, restaurants, and dollar stores—are of limited access to part of the general population due to high prices (though in general prices have been coming down) or because in some cases purchases must be made in dollars, to which not everyone has access.

Attention is given to identifying low-income groups and others with limited access to food at high prices; special programs are designed to assist them. Social assistance covers family income deficits when people are unable to work or their salary is too low. Groups with special needs have programs tailored to assist them. For example, the diets of children, pregnant women, and the elderly are closely monitored. Planning takes into account geographic patterns of distribution of the population, especially with regard to areas of high population density, or limited access, or poor soils, etc. For example, the Turquino-Manatí Plan was designed for the integrated development of four mountainous regions.

Finally, it is important to note that nothing explained here signifies the existence of real poverty in Cuba, which was eliminated through social policy based on the principles of justice and social welfare for all of society. Most analysts agree that the situation in Cuba cannot be adequately evaluated using the normal measures used in other countries—such as the cost of filling the basic market basket—given the variety of different mechanisms of social action that are in place.

Policies, Programs, and Measures Affecting Food Security

NEW ECONOMIC POLICIES

In the new economic policy, beyond measures geared directly towards the agricultural sector, others were adopted to provide food security for the population:

- financial measures to eliminate surplus liquid assets, increase the buying power of the Cuban peso, stimulate the expansion of the economy, and reduce prices, mainly of agricultural products
- the approval and gradual establishment of income taxes regulating income distribution
- the application of new and more efficient management systems in remaining state farms
- legalizing possession of foreign currency, authorizing family remittances from abroad, incentives paid in foreign currency or in convertible Cuban currency, and the development of convertible currency shopping centers, including grocery stores
- revitalization of Regional Food Boards (GTA) presided over by local governments and food producers, to develop plans for sustainable local food self-sufficiency

NATIONAL PROGRAMS

National programs designed to confront food security issues in Cuba can be described in three directions of efforts:

- to provide follow-up and evaluation of the nutritional and food status of the population, and adopt appropriate preventive health measures
- to increase the amount and the nutritional quality of national food production
- to guarantee food availability to all Cubans and provide food and nutrition assistance to vulnerable groups

In the agricultural sector the main programs are as follows:

ESTABLISHMENT OF BASIC UNITS OF COOPERATIVE PRODUCTION (UBPC): A group of measures to support this new form of organization in agriculture is being implemented. These consist of building new housing, promoting self-provisioning and animal traction, and providing training courses on accounting, management, and other production-related topics.

INCREASE RICE PRODUCTION: Emphasis has been placed on increasing yields by using improved varieties, establishing new production methods, soil recovery and improvement, upgrading irrigation and drainage systems, and promoting small-scale rice production.

INCREASE MILK AND BEEF PRODUCTION: Priority on using pasture, sugarcane, king grass (*Pennisetum purpureum*), and legumes for feed—instead of imported concentrates—and on an adequate breeding and production program.

VEGETABLES, ROOTS, AND TUBERS: New varieties have been introduced with faster-germinating seeds, and better management practices are being promoted.

SCIENTIFIC/TECHNICAL PROJECTS FOR SUSTAINABLE FOOD PRODUCTION: Seed production by both traditional methods and tissue culture; plant protection systems; design of models for sustainable development; and other objectives related to the transformation of agricultural production.

MEASURES RELATED TO ACCESS TO FOOD

Measures to guarantee food availability for the population can be grouped into three main areas.

DISTRIBUTION OF STAPLES THROUGH RATION CARDS: The government subsidizes a basic diet, which consists of rice, beans, lard or cooking oil, sugar, a high-protein food (such as beef, poultry, pork, fish, or eggs), bread, flour derivatives, and dairy products. Milk and other products are guaranteed to children up to six years old and to people with chronic illnesses. Havana City and Santiago de Cuba provinces, which have high population densities, have a distribution priority. However, quantities are limited and estimates show that it only covers the nutritional requirements of children from birth to six years old.

SOCIAL DISTRIBUTION OF FOOD: Food is distributed to semi-boarding and boarding schools, day care centers, hospitals, maternity homes and homes for the elderly, as well as to workers' lunchrooms.

SOCIAL ASSISTANCE: A program providing financial assistance to low-income groups under Law 24 of 1979 has four emphases:

- attention to elderly people (financial support and food supply) via community cafeterias near their homes
- social work with single or under-age mothers (financial support for basic food needs, help in obtaining education and employment)
- attention to mentally and physically disabled people (financial support, training in special workshops, and employment)
- attention to under-age children with social problems (those who are out of the educational system or have behavioral problems) or those with families requiring social assistance

The social assistance program has been adapted to the new characteristics of the Cuban society. Financial support has been increased and community participation has been strengthened. In general, these social policy programs have played an important role by preventing further deterioration of food security. This explains why the decline in food security is small relative to the magnitude of economic adjustment.

Nongovernmental Organizations (NGOs)

Cuban NGOs have conceived of and implemented practical, realistic, and less expensive solutions to recover traditional practices of local farmers, promote community participation, and strengthen the capacity of producers to develop innovative ideas and solve problems. Through these activities it has been possible to get more people involved in training, and to promote self-management in food production, as well as build links between local farmers and scientists and professionals who happen to live nearby.

In spite of financial difficulties, Cuban NGOs are promoting projects with food producers that incorporate and guarantee sustainable agriculture principles. These include projects to increase the utilization of renewable energy, make the best use of locally available inputs, improve livestock nutrition and herd management, conserve biodiversity, reduce the use of chemical fertilizers, implement low-input agricultural practices, produce quality seeds, promote the preparation and application of biofertilizers and biopesticides, rescue traditional agricultural practices, and revalue the family farm economy.

A main component of these programs is to promote the role of women in community decision-making with respect to food security. This is a delicate topic in terms of family relations, yet the social role of women is growing and is a significant factor in assuring food security. A large number of rural women have enrolled in secondary schools and higher education in rural sectors, thus attaining a different status in Cuban society. They have become skilled workers, technicians, specialists, and managers in practically all branches of the agricultural sector. Cuban NGO projects are focused on enhancing the role of women by increasing employment possibilities, improving access to information, and promoting a greater participatory role in the family, which has traditionally been male-dominated or patriarchal in character.

NGOs are also carrying out sustainable integrated rural development projects in various regions of the country, especially where government programs have been interrupted due to the financial crisis.

Food security is a number one priority for Cuban NGOs. Among these NGOs are: the Cuban Animal Production Association (ACPA), the National Association of Small Farmers (ANAP), the Council of Churches of Cuba (CIC), the Federation of Cuban Women (FMC), the Cuban Association of Agricultural and Forestry Technicians (ACTAF), the Center for the Study of Inter-American Relations (CIERI), and the Cuban Association of Sugarcane Technicians (ATAC). International cooperation is expected to grow in this area, not just in terms of financial support to specific Cuban NGO projects, but also via the potential for diverse Cuban institutions to collaborate with foreign NGOs in other countries in the region.

Structural Measures

The main structural measures adopted by the Cuban government in agriculture (not including sugarcane production, which has also transformed its productive structure) are as follows:

1. The creation of Basic Units of Cooperative Production (UBPC)

The UBPCs are made up of groups of workers that receive land from former state farms in free usufruct, and buy the means of production from the state. Thus the UBPCs are a new legally recognized form of cooperative farm, and are self-governing according to Cuban law. They are free to make their own decisions. From September 1, 1993 through the first quarter of 1999, 1,612 UBPCs were created with a total area of 1.5 million hectares of land (Table 2).

Table 2: Composition of the UBPC sector, 1999

PRINCIPLE COMMODITY OR ACTIVITY	NUMBER OF UBPCS
Food crops	347
Cattle	719
Citrus and other fruit trees	117
Coffee and cocoa	289
Tobacco	53
Rice	11
Beekeeping	65
Total	1,612

2. New Type State Farms (GENT)

GENTs have been created where the conditions do not lend themselves to creating UBPCs. The main characteristic of the GENTs is that they offer a greater administrative autonomy than did traditional state farms.

3. Distribution of coffee and tobacco land to peasant families

Another important change in agricultural policy has been to hand over land to peasant families willing to produce coffee and tobacco. Although not food crops, coffee and tobacco play a very important role in Cuban economy. Regulations 357/93 and 419/94 of the Ministry of Agriculture dictated the distribution of farm land in usufruct to those who are interested in cultivating them under the proviso that they produce tobacco and coffee as their principal crops, though they are free to grow other crops for family consumption and to sell in the farmers' markets.

4. Distribution of plots for food production

As part of the search for alternatives to increase agricultural production, particularly of foodstuffs, the government has offered up to 0.2 hectares of land in usufruct for family food production with an eye towards self-provisioning, along with permission to sell any surplus. Although the total given out has not been that high, this is another way to increase autonomy in food production and contribute to family food security.

5. Encouraging food self-provisioning

Food shortages forced Cuba to adopt novel ways to meet food requirements. The concept of self-provisioning—growing food for home, community, or institutional consumption—has gradually been implemented throughout Cuba, with emphasis on sectors not specialized in agricultural production.

An important contribution to the nation's food balance has been made, using patches of unused urban land, giving work to the under-employed, using rudimentary tools in many cases, and with an almost total absence of imported inputs, making the resulting product sustainable and organic.

6. Urban agriculture

Urban agriculture is another innovation that has played a significant role in the food supply system of main urban centers in Cuba, especially in providing fresh vegetables and seasonings throughout the year. There are 25 urban agriculture sub-programs, which include livestock production as well as crops. Urban agriculture is organic—no chemical fertilizer or pesticide pollutes the environment or contaminates the food. In 1999 total production reached 850,000 tons, and 1.2 million are projected in the year 2000.

7. Agricultural Production Cooperatives (CPA)

Most of the CPAs were created 20 or 30 years ago and have adapted well to present conditions. These cooperatives were the models for the creation of the UBPCs, since their yields surpassed those of the state farms. The internal organization is similar to that of the UBPCs, but in general these cooperatives are much more consolidated and advanced, and own their land. Many CPA members had been independent small holders before they voluntarily handed over their land to the cooperative. CPAs have more autonomy from the state. CPAs hold 335,900 hectares (not including those dedicated to sugarcane), supplying more than seven percent of total food production in 1998.

8. Credit and Service Cooperatives (CCS)

The CCSs are associations of small landowners who own individual plots, but receive services through their cooperatives, which also help them with credits. Hence, they can take advantage of economies of scale for certain activities. As in the CPAs they were originally small landholders, sharecroppers, and renters, who were then granted ownership of the land on which they worked. Although the majority of CCS members are small farmers, some have reasonably large areas they farm. There are 168,000 members of CCSs, with 979,900 hectares.

A campesino *farm: small-scale, diversified food production.*

Future Measures

The past 10 years of economic development have demonstrated the ability of the Cuban people to *resist*, while carrying out economic transformations to consolidate their progress and overcome difficulties. A number of important tasks are still pending, though initial steps have been taken:

- The highest priority will be placed on any and all measures to support agriculture and food production.
- The state will strengthen its active role in the economy while continuing to develop new organizational forms and a diverse array of cooperative, private, and mixed forms of property.
- Reviving the sugar sector will have a top priority.
- The process of creating UPBCs will continue.
- The Regional Food Boards will play a key role in establishing agricultural policies and in making strategic decisions for local food availability.

- Those resources that are available will be mobilized, and production incentives will be used, to stimulate national food production and further reduce the need for imports.
- Restructuring and modernization of the financial sector will continue, with emphasis on institutions that provide credits for the development, production, and marketing of agricultural products.
- The resurgence of the food industry, taking into consideration available resources, cost minimization, and technological and energy efficiency, will be encouraged.
- New mechanisms for regulating the economy will be developed, with an eye toward balancing the new reality with the planned nature of the Cuban economy.
- The business sector will be redimensioned, and an alternative employment program will be established.
- Social policy will be adapted to the new characteristics of income distribution in Cuba, with greater assistance to more vulnerable sectors of society.
- The overall goal of change will remain the perfecting of the socialist system, while balancing the introduction of new market mechanisms with sustained planning via foresight and anticipation of needs.

CHAPTER FOUR

Transforming the Cuban Countryside: Property, Markets, and Technological Change

Lucy Martín, Center for Psychological and Sociological Research

The present situation in Cuba's countryside is as interesting as it is complex. It is characterized by the reformulation of strategies to permit adjustment to new circumstances, and a slow emergence from the economic crisis, without giving up the basic principles that define the Cuban social project. From a social and structural perspective, the following questions arise: What type of class structure is being formed in agriculture? What elements differentiate the main groups? What does this emerging structure contribute to the social model that is emerging? And does this structure fit with the Cuban model of equity and social justice?

Of course I do not have the final answers to these questions, but any attempt to address them begins with a basic understanding of the main processes coming together in the new model of agricultural development. We need such an understanding to identify the principal scenarios (according to property types) where these changes are taking place, and to tease out the characteristics of production of the different sectors associated with each of the types of land tenure. This chapter will address the production characteristics of the most important agricultural social groupings—those who are entrusted with carrying out today's transformations, and without whose active and conscious participation it would not be possible to achieve a process of change leading to more efficient and rational production systems.

The Fundamental Processes: A Necessary Overview

The transformations in Cuban agriculture, seeking to overcome the economic crisis through a more endogenous development model, imply profound changes in technological and economic policy. These changes are leading us away from the conventional agricultural model, which was distinguished by dependence on oil and chemical inputs, large extensions of state land with salaried workers, and a highly centralized planning and management process. The new policy directions emphasize sustainability, with the following essential features:

- the co-existence of different systems of property and management
- the combination of market and centrally-planned economic mechanisms
- progressive modification of agronomic practices
- technological change based on sustainable agriculture
- the development of a more participatory economy with a tighter linkage between workers and farmers and the profits of their labor

It should be noted that during the period preceding the economic crisis, food production showed sustained growth for almost every commodity,[1] although production levels did not fully cover the needs of the population and much food was imported. In addition, production levels did not truly reflect the large investments made in agriculture. Yet, the infrastructure built by the revolutionary government produced one of the most highly technified agricultural systems in the hemisphere.

During this period state farms were the most important producers of staple foods such as rice, milk, beef, poultry, roots, and tubers, and were the main exporters of sugar, citrus, and coffee. Private farmers produced most of the beans, maize, vegetables, tobacco, and cocoa.[2]

In such a retrospective analysis the issue of production efficiency in each sector must be considered. Small farmers were the minority in both employment and total acreage, and in general were less technified and had fewer material resources. Nevertheless, the productivity of private farmers was greater than that of state farms in certain key commodities, and they reached the same or higher yields with fewer inputs. With only 20 percent of the total agricultural area, private farmers contributed 35 percent of national production, using less than 20 percent of the resources invested in agriculture (Nova 1994).

In the 1990s, there were two important factors in the growth of the *campesino* or non-state sector: a massive expansion of land area under private and/or cooperative production, and the reopening of the farmers' markets where surplus production is sold at free market prices. In what follows I reflect upon the impacts of the principle transformations of the last five years, in terms of property, market mechanisms, technological change, and participation.

Restructuring Land Tenure and Management

Since the end of 1993 important modifications of land tenure have been made, and while not establishing new production relationships, they have provided the necessary conditions to boost economic effectiveness. These were the first major changes since 1977. Many of the state farms have been broken up into smaller scale farms called Basic Units of Cooperative Production (UBPCs), and the agricultural sector was opened to foreign investment in joint ventures with the state. Furthermore, unused lands were distributed in usufruct to new farmers. These changes represent a broadening of the land tenure matrix, generating a new mixed economy which is based primarily on individual farmers—with both private and usufruct tenure—, private cooperatives, and collective farms with usufruct tenure (Figueroa 1996). We can identify ten distinct forms of organization that can be grouped into three sectors (see Table 1).

Table 1. New forms of organization

<table>
<tr><td colspan="2" rowspan="4">STATE SECTOR</td><td>State farms</td></tr>
<tr><td>New Type State Farms (GENT)</td></tr>
<tr><td>Revolutionary Armed Forces (FAR) farms, including farms of the Young Workers' Army (EJT) and the Ministry of the Interior (MININT)</td></tr>
<tr><td>Self-provisioning farms at workplaces and public institutions</td></tr>
<tr><td rowspan="5">NON-STATE SECTOR</td><td rowspan="2">Collective Production</td><td>Basic Units of Cooperative Production (UBPC)</td></tr>
<tr><td>Agricultural Production Cooperatives (CPA)</td></tr>
<tr><td rowspan="3">Individual Production</td><td>Credit and Service Cooperatives (CCS)</td></tr>
<tr><td>Individual farmers, in usufruct</td></tr>
<tr><td>Individual farmers, private property</td></tr>
<tr><td colspan="2">MIXED SECTOR</td><td>Joint ventures between the state and foreign capital</td></tr>
</table>

The State Sector

From the beginning of the Revolution through the early 1990s, the state was the most important sector of production. In recent years, however, the state farms have been drastically downsized in terms of landholdings, number of workers, and equipment, reducing the economic importance of the state sector. Since 1993 holdings in the state sector have shrunk from more than 75 percent of the arable land to less than 33 percent in 1996. The remaining workers in the state farm system are concentrated in strategic fields such as animal breeding, large-scale pig and poultry production, and others that require heavy mechanization, qualified personnel, and emerging technologies. Thus most of the remaining state farm employees have little in common with traditional farm workers.

The state sector also includes farms belonging to the Ministry of Interior (MININT), the Revolutionary Armed Forces (FAR), and the Young Workers' Army (EJT)—a subdivision of the FAR. The primary purpose of these production units is to provision their respective organizations, though they also sell considerable quantities of surplus food through state-owned wholesalers. The EJT has its own commercial arrangements and is one of the most efficient producers within the state agricultural sector.

Self-provisioning areas are also considered part of the state sector. During the 1990s many factories and workplaces, and public institutions like hospitals, schools, and office buildings, were given unused lands to produce food for their own cafeterias, and to sell to their employees at reduced prices. Today subsidized lunches and discount produce represent an important non-salary benefit for Cuban workers.

The New Type State Farms (GENT) are the final component of this sector. After the creation of the UBCPs it became evident that the transition from state farm worker to farmer was one that could be made easier if progressive transitions were made. The GENTs are completely owned by the state, but worker cooperatives are built upon them, and over time they take on more financial and management responsibilities. At a minimum, they enter into profit-sharing schemes with the underlying state farm structure. Rather than being state employees, the cooperative members enter into a contract with the state and the cooperative's profits are shared among the workers according to their own internal agreements. In the GENTs, both profit and risk are shared between the state farm and the worker cooperative, but minimum salaries are guaranteed, while ultimate responsibility for the farm and key management decisions are taken at the state enterprise level. There is a great deal of flexibility in these experimental arrangements, allowing each division and even particular enterprises and farms to work out their own arrangements within certain parameters. The final destiny of a given GENT might be the creation of a UBPC, or it might not.

The Non-State Sector: Collective Farming

In the non-state structure there are two principle forms of production. The largest is collective production, where the land is worked jointly by all cooperative members, and management decisions are made through democratic processes. In individual production, each farmer's plot is worked basically on a family farm model. Most of these farmers are also members of cooperatives, to access services and credit, to purchase inputs in bulk, and to sell their produce, though production itself remains individual. Within this sector there are also two main types of land tenure, private and usufruct, which cut across both forms of production.

AGRICULTURAL PRODUCTION COOPERATIVES (CPA): The CPAs are the traditional revolutionary form of cooperative production in Cuba, founded in 1977 by farmers who voluntarily choose to unite their private individual lands and resources for increased production, marketing, and economic efficiency. In 1997 there were 1,156 CPAs with a total of 62,155 members, who owned 9.4 percent of the agricultural lands (ONE 1997). The CPAs had shown a steady decline in membership from the mid-1980s to the early 1990s, when they began to rebound. The recovery came about as new members joined, with backgrounds in the most diverse array of occupations. They were drawn to farming by the advantages of rural cooperative life with respect to income, access to affordable food and, to a lesser degree, housing.

Because many of these new members come from other fields that surely have different styles of workplace discipline, habits, and motivations, one can assume they need a period of adjustment and training. Although it is but a working hypothesis at this time, it is likely that as a result of this influx of new blood, the present membership of CPAs is now more heterogeneous in respect to social origin, professional characteristics, and needs and interests. At the end of the 1980s the average age of a cooperative member was 41 years old, but it is likely that the growth in numbers of new members has brought down the average in recent years.

BASIC UNITS OF COOPERATIVE PRODUCTION (UBPC): The CPAs were joined by the UBPCs in the early 1990s. With the creation of the UBPCs at the end of 1993, a new type of cooperative was established, not by the voluntary socialization of private property, but rather through the de-statization of state property and infrastructure.

The UBPCs are productive units with a cooperative structure, that farm state lands which were given free of charge to the cooperatives in permanent usufruct (the average acreage is substantially smaller than the former state farms, which have been broken up to form the UBPCs). Other means of production such as buildings, machinery, animals, irrigation systems, and tools, were sold to the cooperatives at favorable prices with low-interest loans, and as such constitute private property of the cooperative. The UBPCs maintain commercial relationships with the distribution chain of the original state enterprise from which they emerged, and negotiate prices and production plans based on a quota system. Surplus production is sold at the farmers' markets

Campesino *family at the "28 de septiembre" farming cooperative, Batabanó municipality, Havana.*

at prices set by supply and demand, and through other outlets. The UBPCs also receive technical support from the enterprise, from which they purchase inputs and additional equipment as needed.

In terms of numbers and area the UBPCs are now the predominant type of farm in Cuban agriculture. In 1997 there were 2,654 UBPCs with 272,407 members occupying 42 percent of the land (ONE 1997). By 1995—after less than two years of existence)—23 percent of the sugarcane producing UBPCs and 52 percent of the non-sugarcane producing UBPCs were profitable. This is a vast improvement over the situation among state farms prior to 1993 (Rodríguez 1996).

The new cooperative members, who were the workers of the previous state farms, constitute a new type of producer who must face the challenge of achieving greater economic efficiency. This social and structural transition is currently underway, and in some cases there continues to be a certain degree of ambiguity between the previous structures of state-run enterprises and the new cooperative structures. Additionally there is a psychosocial transformation underway in which the agricultural worker must transcend his or her previous function and mindset and become a true cooperative owner-operator—in other words, a former farm worker must become a farmer, not always an easy transition.

With the appearance of the UBPCs a new economic player has emerged in Cuban agriculture: the cooperative farmer on state lands. These farmers are now the most important sector with the largest number of people involved.

They have the dual responsibility, or social duty, of achieving higher levels of production, but with fewer inputs and other resources. To do this they must break with the conventional ways of doing agriculture that were established in the former state farm sector. In some cases, the relationships between the new cooperatives and the old state enterprises have been marked by an excess of tutelage, subordination, and dependence, remnants or legacies of an enterprise management structure that has not yet fully given way to a more appropriate and participatory planning process among actors. The UBPCs demand a great deal of attention and support because they make up a new and very large grouping, which must play a key role in the new national production strategy.

Analysis of these two forms of production show that the cooperative sector as a whole has flexibility, heterogeneity, the ability to combine diverse crops and technologies, a qualified labor force, and an unquestionable capacity to form groups with common interests (economy, ideology, community, and even family interests). These factors combined with the large acreage, sheer number of members, and social responsibility, make it the most important part of the new social structure of Cuban agriculture.

The Non-State Sector: Individual Farming

Individual small farmers who work their land based on a family farm model can be classified into three major categories. Most of those who have private ownership of their farms are members of Credit and Service Cooperatives (CCS); then there are the individual farmers who have received lands in usufruct from the state in recent years; and finally, dispersed individual farmers who are not coop members. After a sustained decrease in numbers in the 1980s, in the 1990s the individual farmer sector began to recover both in terms of numbers and acreage. Today they hold 55 percent of the private farmland in Cuba—up from 42 percent in 1988—and thus are economically important. While there are some dispersed individual small farmers, the majority of farmers producing in individual farms are members of CCSs.

CREDIT AND SERVICE COOPERATIVES (CCS): In 1997, there were 2,709 CCSs, with a membership consisting of 159,223 individual farmers working 11.8 percent of total agricultural land (ONE 1997). In this type of cooperative, individual farmers work their farms independently, but join together to receive credit and services from state agencies. They may also share certain machinery and equipment, especially in the so-called "Strengthened CCSs." In any case the land continues to be individual property, independently managed by the owner. CCS members purchase inputs and sell products at fixed prices through state agencies, based on production plans and contracts established with state distribution systems. Any production above and beyond the contracted quantity may be sold in the farmers' markets at free market prices.

During the recent period there has been an accelerated growth in the numbers of new farmers in the CCSs, even more so than in the CPAs. This may be explained by a number of factors, but in essence it is an economic phenomenon. It comes down to the fact that individual farmers have higher incomes than do members of production cooperatives. Perhaps this is because they are able to make faster decisions and because they have a greater sense of ownership, or because their management practices lead to more efficient use of limited resources. Possibly it is because they have less of a sense of social responsibility or, more likely, it is due to a combination of these factors. As a rapidly growing group of farmers, they too have undergone a demographic shift, as most of the new members are young, lowering the average age, which was about 50 in the 1980s (Domínguez 1990).

INDIVIDUAL USUFRUCT FARMERS: Beginning in 1993 individual families were given up to 27 hectares of land in free and permanent usufruct to grow specialty crops such as coffee, tobacco, and cocoa. By 1996 the number of these so-called *usufructuarios* had grown from zero to 43,015 farmers (Lage 1996). In addition to this group, in many urban areas individuals were given small plots of land (0.25 hectares) to grow food for themselves and their neighbors. These new farmers come from diverse backgrounds, although it is likely that most of them were previously workers or professionals in agricultural fields. The National Association of Small Farmers (ANAP) aims to incorporate these new farming families into the CCSs and into the Association.

The Mixed Sector

Joint venture enterprises with foreign companies exist in the citrus export industry, and some other export commodities have received foreign financing for a portion of national production (rice, cotton, tomatoes), which may expand to other crops in the near future. How the agricultural labor force will be affected by this, and what traits will identify and differentiate the new group of workers linked to foreign capital are topics for future research. It should be noted, however that only state enterprises can accept and use foreign capital. Thus no private producer(s) can establish direct relations with foreign investors. This measure has been implemented by the state to regulate the sort of social and economic differentiation that might otherwise arise from these activities.

The analysis of these very different forms of organizing production, and the different social groups associated with them, shows a great socioeconomic and structural diversity in contemporary Cuban agriculture. Still, and this is very important, it is precisely this heterogeneity—conceived of as part of an articulated system—that allows for the application of distinct and varied technological alternatives. Each of the forms described above has particular characteristics to offer, that when integrated at a system level, could represent greater strength and integration in the system as a whole. Through a network of connections and relationships, through which they can interact and compliment one another, both collaboration and competition can be stimulated.

Commercial Flexibility and the Introduction of Market Mechanisms

Historically, distribution and marketing of agricultural commodities to the whole population at accessible prices has been the responsibility of the state, except for a brief period (1980–1986) when the earlier experiment with free market farmers' markets was carried out. With the opening of the farmers' markets in 1994, a step was taken toward optimizing production relationships, allowing surplus production—above and beyond amounts contracted for with the state—to be sold at free market prices based on supply and demand.

The ability to get higher prices and raise incomes by surpassing contracted production quotas has led to a more active and efficient management of productive resources, with one outcome being the greater availability of food for the population. 1996 sales data from the farmers' markets reveals the dominance of individual producers in this venue (see Figure 1).

Figure 1. Sales at farmers' markets, 1996

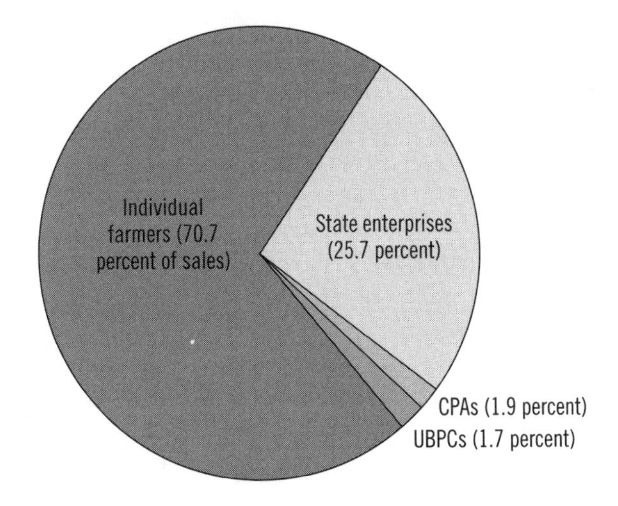

(ONE 1997)

Although the new markets have had repercussions for all producers, this change is quite significant for the former farm workers/state employees who are now either members of UBPCs, or of cooperatives within GENTs. For the first time they have the opportunity to make extra income through collective self-management. While they currently participate to a very limited degree only in the markets, there is great potential for this sector to participate more actively. The act of bringing their own production to the marketplace helps bridge the gap between producers and consumers, offers them more freedom of action, and greater individual and collective incentives.

Central planning and market mechanisms are not mutually exclusive. The latter has its specific role to play—even when its negative impacts on social equity are recognized—as it currently does in the evolution from central planning toward worker self-management. The tension between the two is expressed as a necessary equilibrium. This delicate balance must simultaneously promote initiative, reduce alienation, and provide economic incentives, such that agricultural production increases—without permitting a fall into anarchy, ungovernability, or the loss of the state's ability to maintain and meet the key social objectives of socialism. In concrete terms, nonsocialist forms of production must have a subordinate role in which there is a certain level of control over windfall private profits. In the final analysis, the social and structural expression of this transformation can be described as a socioeconomic strengthening of the agricultural workforce, internally differentiated by property type and management system, and modified by the particulars of commodities produced, geographic areas, distance from markets, and availability of transport.[3]

Participation and Initiative

The ideal of a state production system based on high-tech methods and centralized decisions was confronted with the reality of practices involving excessive use of resources and inputs—well beyond the point of diminishing returns—, underutilization of expensive investments in infrastructure, a low degree of agility in decision-making and provision of services, and the adoption of production norms far removed from real needs. Yet the need for transformation ran deeper still.

As we have said, the changes carried out to date foster, at least theoretically, enhanced participation, more taking of initiative, reduced alienation, and greater realization of the human potential of workers and farmers, though the effects to date are as yet uneven. To illustrate the impact of these changes on producers, we refer to their own varied experiences:[4]

A woman who is president of a UBPC told us: ". . . when I was the head of the state farm I did not think as I do now. Only when one is here does one really know what things are really necessary, and you try to spend as little as possible so that the cooperative can be profitable." She went on to give an example, of when ". . . the state enterprise charged us for trucking in water twice a day and this cost us almost 30 pesos. . . and I said to myself, this cannot go on any longer. . . so I jumped on a tractor, hooked a water tank on a trailer to it, and now the whole operation doesn't cost us even two pesos."

A woman at a UBPC dairy unit was asked if she worked according to a strict, predetermined schedule. She responded, "I don't even know how to answer you; no, we don't have a schedule—if I have to bathe the animals I do it until I've finished. At the end of the day we're the ones who have to do what has to be done. I can't say I am going to quit now because it's time to go home, or think, oh, I'll let that other guy do it." Later she added: ". . . we have to worry even more about the work because since my husband is the administrator I have to help him out."

A different view was expressed by the manager of a UBPC: ". . . a *campesino* does not understand the constraints limiting development possibilities. . . he only understands what he has in his pocket. . . today, *campesinos* participate more because they are under new social conditions, but they will only feel a sense of ownership when they see the benefits of their work, and this will be when there are profits—and I do not see that we will make profits in the near future."

A UBPC member expressed that, ". . . regardless of the fact that more efficient land use is achieved when sweet potato and maize are intercropped, in the past provincial officials didn't allow us to grow these crops together. . . though everybody knows that *campesinos* have always done this. Even today they become uneasy when we say we don't need to apply chemical formulas to soils when planting potatoes, because there is enough phosphorus and potassium available. When we applied chemical fertilizer to potatoes, the yields were no higher than when we did not. . ."

This situation should evolve toward new forms of management and the establishment of more productive relationships among production units and between these and state enterprises. This should allow the emergence of and development of production units which exercise their full rights and take advantage of the will and creativity of their members. The key is that the link with the state should not be one of dependence and subordination, but rather a two-way transmission between different mechanisms and centralized state planning.[5]

Technological Change

In parallel with social and structural transformations, Cuban agriculture is undergoing change from a conventional, highly technological model, toward an agriculture based on appropriate technologies. Cuba had embraced intensive agriculture, with the accelerated introduction of high-tech methods and chemical inputs in search of higher yields and greater labor productivity. But our agriculture eventually had to confront the declining response of crops to ever higher rates of chemical fertilization, worsening soil degradation (compaction, salinization, and erosion), growing resistance of pests to pesticides, more frequent pest outbreaks as natural mechanisms of regulation broke down, an overemphasis on extensive monoculture, and higher wind speeds and other climatic changes due to deforestation, among the many negative effects of industrial agriculture. To these powerful reasons for change can be added the simple impossibility of maintaining high levels of imports of chemical inputs, fuel, and machinery in the wake of the collapse of trade with the former Socialist Bloc.

A no less important factor driving technological change was the impact in low productivity of the work relations inherent in the dominant state-led production model. In the name of intensifying agricultural production, priority was given to state farms, production was largely placed in the hands of salaried farm workers, and highly centralized management practices were employed in both production and distribution.

All of these factors came together to drive technological change in search of higher productivity. The new model, however, cannot be simply reduced to its technical dimension—agroecological management of production systems—, as is often the case with technological change in other parts of the world. Rather the technical dimension is intimately entwined with something as fundamental as changes in the relations of production, which makes the change process in Cuba a more integral one, and differentiates it from experiences elsewhere.

Among the elements that Cuba has mobilized to carry out this transition is science policy aimed at the national development of new technologies. This lays the groundwork for a more fruitful relationship between technology and development, and allows us to make changes—with our own identity—to bring about more sustainable development. This is not without its apparent contradictions, however.

Since the mid–1970s diverse actions have been taken which tend toward raising production in consonance with the environment. These steps include national production of improved seeds, biopesticides and biofertilizers, the growing use of biological pest control, and improved methods of soil conservation, etc. Yet this national process of applied research has as its maximum expression the use of agricultural biotechnology, which poses the risk of altering life as we know it. Both movements, one toward an agriculture more in harmony with nature, and the other which tries to separate agriculture from the natural environment, are found in Cuba.

The nature of a given technology and its degree of introduction will produce unequal effects depending upon the characteristics of the environment in which it is inserted and, above all, on the particular social actors in charge of its implementation. Thus the search for appropriate technologies should contemplate the coexistence of cutting-edge technologies such as embryo transfer with the most rudimentary ones such as animal traction, so as to address the mix of needs and potentials of production and producers.

In the world today—including Cuba—there is an ongoing debate between schools of thought that give greater weight to the technological paradigm of modern agriculture and those who promote organic farming (or alternative, or natural agriculture). However, this contradiction loses much of its meaning, at least in Cuba, if we take into consideration the great diversity of forms of property, soil conditions, environments, and so on. Taking this great heterogeneity as our starting point, it makes more sense to talk about adapting appropriate technologies for sustainable agriculture, with room for considering all technological alternatives, all kinds of inputs and equipment, to satisfy varied local requirements, an approach which we might call "technological pluralism."

The strategies we adopt must be flexible and adaptable to different local realities, where elements of traditional and of conventional high-tech paradigms may coexist. Even the most rudimentary traditional methods should not be seen as limitations imposed by economic exigencies, but rather as elements in a technological diversity that corresponds to the socio-structural heterogeneity of Cuban agriculture.

The success of a technological strategy depends on rural extension work, where the active subject of change is the worker/farmer who implements the new technologies. Proposals must be elaborated, and plot-by-plot feasibility assessed, hand in hand with the farmer, in order to achieve a better fit between technology (equipment, inputs, agronomic practices), varied forms of land tenure, and local environmental characteristics.[6]

Conclusion

We can conclude that the transformations described in this chapter are leading toward a strong social and economical differentiation of the agricultural labor force, accentuating already existing social and structural heterogeneity and the complexity of the very processes of change. From a social and structural point of view, we perceive a broadening of the agricultural labor force and an increase in its internal heterogeneity, which can be expressed in the following:

- the emergence of new social agents in agriculture: cooperatives on state lands, *campesinos* with lands in usufruct (individuals and family groups), and workers in joint venture enterprises

- an increase in the size and diversity of the agricultural work force
- the greater numerical and economic importance of *campesinos* in general, and of those who own their land in particular
- the numerical predominance of cooperative members
- specialization and reduction in the overall number of less skilled farm laborers
- greater social differentiation within the agricultural labor force, driven by the new diversity in the ways in which production is organized, and the variation in market linkages and technologies being used. This differentiation is expressed in terms of income, living standards, and the needs and interests expressed

The form of property through which they access land, the fundamental economic activity to which different groups are connected, the variation in production resources to which they have access, and incomes and housing conditions all continue to carry weight as factors of differentiation. But as never before, *the* mediating or determining factor is participation in the market. This can be seen clearly in the diminished capacity of the state to regulate the process of social differentiation among farmers. Greater and greater weight is accruing to the organizational forms and methods of resource management that each farmer or farmer organization can mobilize and bring to the table. Strongly related to this is the fact that economic incentives, reduced alienation, and greater efficiency in using scare resources emerge strongly as key factors in all of the evolving scenarios in Cuban agriculture—though of course with great variation across organization forms.

The fragmentation of property, opening of markets, and technological change, all have a decentralizing nature and tend to strengthen local economies. This also favors a decentralization of production in the countryside. Processes of fragmentation are occurring simultaneously with processes of socioeconomic strengthening, but—in contradistinction to other realities—all of this is taking place in the context of a logic and a society in which the state plays the role of protagonist in redistributing wealth to the majority.

Notes

1. In a press interview Carlos Lage (secretary of the Executive Committee of the Council of Ministers) confronted the suggestion that state agriculture was a failure. He pointed out that since the triumph of the Revolution in 1959, overall agricultural production increased many times faster than population, in some cases reaching levels of efficiency rivaling other countries (*Granma*, October 30, 1993, pg. 6).

2. In 1999, according to the National Office of Statistics, the *campesino* sector produced 86.3 percent of tobacco, 68 percent of maize, 72.6 percent of beans, 46.6 percent of roots and tubers, and 55.9 percent of cocoa.

3. Now more than ever, socioeconomic differentiation has deepened among the *campesinos*, and it is more evident due to the differentiating role of the market. In a recent analysis by the National Association of Small Farmers (*Caracterización de los actores sociales campesino y cooperativista. Esfera agroalimentaria,* 1996), it was noted that market insertion guarantees higher incomes for those cultivating basic food crops, while the lowest incomes were recorded by those producing coffee, cocoa, and fruits in the highlands since they have less access to the market, because of the amounts they are authorized to sell, and because of transportation difficulties.

4. This and the following testimonies were collected in a field research study, *Presencia femenina en las UBPC ¿Un problema aún por resolver?,* by Teresita Almaguer et al. Centro de Documentación, Federación de Mujeres Cubanas (FMC), 1996 (mimeo).

5. In this respect the analysis and proposal contained in a report by José Luis Martín and Angel M. Suero are very interesting. "La competencia decisional en las UBPC," *Agroecología y Agricultura Sostenible. Curso para Diploma de Post-Grado.* Módulo 3. CLADES-ISCAH. 1997. See also a report by Dr. Miguel Limia entitled "Las UBPC como forma embrionaria de un nuevo colectivo laboral," *Resumenes de investigaciones sobre UBPC.* Programa FLACSO-UH, 1994.

6. An experiment in local technological change is being carried out by a multidisciplinary team linked to the UNDP-SANE project in three CPAs producing food crops in Havana province. See Eolia Treto et al., *Informe de Investigación.* Instituto Nacional de Ciencias Agrícolas (INCA), 1998 (mimeo).

References

Domínguez, M.I. 1990. *Diferencias y relaciones generacionales en el campesinado.* Havana: Centro de Investgaciones Psicologicas y Sociológicas.

Figueroa, V. 1996. El nuevo modelo agrario en Cuba bajo los marcos de la Reforma Económica. *UBPC, Desarrollo rural y participación.* Havana: Colectivo de Autores.

Lage, C. 1996. Informe al IV Pleno del Partido Communista de Cuba. *Granma* 26:3: 1996.

Nova, A. 1994. Cuba: modificación o transformación agrícola. INIE (mimeo).

ONE. 1995. *Estadísticas agropecuarias. Indicadores socioles y demográficos de Cuba.* Havana: Oficina Nacional de Estadísticas.

ONE. 1997. *Estadísticas agropecuarias. Indicadores socioles y demográficos de Cuba.* Havana: Oficina Nacional de Estadísticas.

Rodríguez, J. 1996. Cuba 1990–1995. Reflexiones subre una política económica acertada. *Revista Cuba Socialista* 1:24.

CHAPTER FIVE

Social Organization and Sustainability of Small Farm Agriculture in Cuba

Mavis D. Álvarez, National Association of Small Farmers

Very little is known outside of Cuba about the social organization of Cuban farmers. Even experts and people familiar with farming are frequently confused by private and cooperative agriculture in Cuba. In both public presentations and personal meetings, I frequently witness surprise or doubt regarding the existence of private farmers in Cuba. It is clear that this is often due to preconceptions and prejudices about processes such as nationalization, state holdings, socialism, and cooperative structures.

I don't intend to clarify theoretical concepts, which would be impossible in a short piece such as this. It will be much more useful to examine in the greatest possible detail the real-world elements of rural Cuba, as formed by private and cooperative farmers, and a perhaps more interesting and polemical topic: the differentiation between private and cooperative properties.

The central questions in this discussion are: Do Cuban cooperatives really represent a "private" model of agriculture? How are private and cooperative production organized? What role do these sectors play in the national agricultural economy and food supply? Has the significant transformation of the Cuban economy in the last decade recognized the importance of these sectors? And finally, how have small farmers participated in the development of a national sustainable agricultural model?

Cooperatives and Small Farm Owners: Social Organization of Production

In Cuba the so-called "non-state agricultural sector" is comprised of the newly formed Basic Units of Cooperative Production (UBPC), the older Agricultural Production Cooperatives (CPA) and Credit and Services Cooperatives (CCS), and a minority of small landowners. The latter, called *parceleros*, are dispersed throughout the countryside and are not associated with any of the existing cooperatives. They work their small parcels of land primarily for family consumption and sell their surplus in local markets.

The "non-state sector" is a conventional term used to define production units which have legal right to the land, as private property (CPA and CCS) or in usufruct (UBPC, parcel owners, and others). These units of production differ from other agricultural entities situated on public lands where property rights rest solely in the hands of the Cuban government.

Table 1: Number of cooperatives by type and province

PROVINCE	CPAS	CCSS	TOTAL
Pinar del Río	125	359	484
Havana	64	219	283
City of Havana	1	47	48
Matanzas	71	94	165
Villa Clara	93	268	361
Cienfuegos	41	67	108
Sancti Spíritus	62	143	205
Ciego de Ávila	60	75	135
Camagüey	98	138	236
Las Tunas	57	152	209
Holguín	128	299	427
Granma	128	218	346
Santiago de Cuba	106	235	341
Guantánamo	102	256	358
Isle of Youth	3	8	11
National Total	1,139	2,578	3,717
Total %	31%	69%	100%

(ANAP 1998)

The non-state sector is comprised primarily of CPAs and CCSs (see Table 1). Together CPAs and CCSs occupy 22 percent of the agricultural acreage of the country: more than 1.5 million hectares of land apt for economically viable agricultural production. These types of cooperatives, and the individual small farmers, will be the primary focus of this chapter.

Tobacco cultivation with animal traction at a campesino *farm.*

Table 2. Number of hectares owned by cooperatives

PROVINCE	CPAS	CCSS	TOTAL
Pinar del Río	50,906	151,178	202,084
Havana	52,831	58,195	111,026
City of Havana	282	5,027	5,309
Matanza	82,925	44,640	127,565
Villa Clara	69,518	86,113	155,631
Cienfuegos	26,785	32,403	59,188
Santi Spíritus	47,894	69,496	117,390
Ciego de Ávila	57,832	35,343	93,175
Camagüey	84,269	68,609	152,878
Las Tunas	37,243	47,992	85,236
Holguín	65,606	100,659	166,265
Granma	47,423	87,738	135,161
Santiago de Cuba	52,605	96,810	149,415
Guantánamo	32,238	91,754	123,992
Isle of Youth	1,586	3,930	5,517
National total	709,944	979,888	1,689,832
Total %	42%	58%	100%

(ANAP 1998)

Most of the lands now under CPAs and CCSs were distributed to farmers shortly after the Cuban Revolution, during the first and second agrarian reforms of May 17, 1959 and October 3, 1963. The goal of these reforms was to lay the groundwork for future rural development by ensuring a livelihood for the rural population. Unlike less successful agrarian reforms in other parts of the world, Cuba's agrarian reform laws required that land given as property to small farmers be productive and capable of supplying foods and financial security to owners and their families (La O Sosa 1997). The laws even allowed for exchanges and relocations in order to offer the new owners better living and working conditions. Property rights included not only land, but also the materials necessary for production, such as farming implements, plows, housing and other buildings, as well as ownership over the harvest itself. These laws are still in place and all subsequent agrarian legislation respects and observes these rights.

Table 3: CPS and CCS membership

PROVINCE	CPAs	CCSs	TOTAL
Pinar del Río	6,791	26,707	33,498
Havana	5,887	17,579	23,466
City of Havana	25	3,313	3,338
Matanzas	6,492	7,281	13,773
Villa Clara	5,380	20,790	26,170
Cienfuegos	1,808	8,711	10,519
Sancti Spíritus	4,842	13,212	18,054
Ciego de Ávila	4,870	5,148	10,018
Camagüey	4,499	6,722	11,221
Las Tunas	3,128	6,000	9,128
Holguín	5,618	15,983	21,601
Granma	5,029	10,621	15,650
Santiago de Cuba	5,744	13,204	18,948
Guantánamo	2,740	12,842	15,582
Isle of Youth	72	371	443
National total	62,925	168,484	231,409
Percentage	27%	73%	100%

(ANAP 1998)

Agricultural Production Cooperatives (CPA)

Members of the CPAs are private property owners—CPAs are *not* public land. These cooperatives are defined by the transformation of individual property into social or collective property. A CPA is the voluntary association of small farm holders who combine their interests and material resources (land and equipment) to create a common holding that is administered by all, and is worked for the benefit and development of the entire community. The CPA is an economic and social organization, and its management is autonomous from the state, it has its own legal status and carries out its own activities in the best interests of society at large, and according to internal cooperative democracy and the collective work of its members.

The cooperative is the legal owner of all of its productive resources including land, equipment, housing, animals, crops, forests, and other natural resources, and any other assets purchased by the cooperative or contributed by its members. Cooperatives also have full rights to the profits and financial reserves generated by their production. They can increase their financial and material holdings through the work and contributions of individual members.

The goals of a CPA include:

- agricultural production that serves the best interests of society and of the cooperative
- to consolidate and increase the social benefits from production within the CPA
- to increase production of marketable agricultural goods
- to ensure labor productivity and the efficiency of communal production
- to promote appropriate science and technology
- to meet the material and cultural needs of cooperative members and their families and improve the general welfare of the community
- to encourage participation in diverse aspects of community life and enhance coexistence among members

To ensure that the CPA will be governed by a standard of cooperative democracy, the Agricultural Cooperative Law states that the independent legal status of a CPA is based on a General Membership Assembly. All cooperative members participate in the assembly, making it the highest administrative authority in a CPA. It elects a board of directors, which conducts executive and management duties, though the board is directly accountable to the assembly. The assembly is able to elect new board members and to renew and remove individuals from their posts. The president of the board directs the activity of the cooperative and implements decisions made by the board, or by the full General Assembly.

The General Regulation for CPAs, a measure approved by the Council of Ministers in Decree No. 159 on September 20, 1990, regulates the duties and powers of the president and the board of directors and delineates the rights and responsibilities of cooperative members. All members of a CPA have the right to share in its profits according to the quantity and quality of their work. Throughout the year cooperative members receive payments based on the number of days worked, and, once the annual financial analysis is complete, remaining profits are divided among the members. Profits from sales can also be placed in communal funds to meet specific needs. The General Assembly is responsible for deciding which funds should be created, how they can be used, and the percentage of gross income dedicated to each fund.

Credit and Service Cooperatives (CCS)

The Agricultural Cooperative Law defines a CCS as a voluntary association of independent small farm holders, for mutual economic support. Members own their individual assets (property, equipment, and production profits), and work their own land. Children, parents, siblings, or a surviving mate may inherit these lands provided they have been present and contributing to the farm over the previous five years. This model is consistent with the traditional definition of private ownership: the farmer works his or her own land for his or her personal profit. However, the farmer is also economically integrated into the community through membership in a CCS.

CCSs are organized to promote cooperation by allowing farmers and their families to work together to improve their financial and social conditions. Together the cooperative members are better able to take advantage of economic and technical services and material resources. CCSs can request state assistance, both financial and technical, through their annual Technical and Financial Plan that describes the real and potential human and material resources of all cooperative members. The CCS must then coordinate the disbursement of state assistance among its members. The CCS also contracts non-state technical and material inputs for all member farms, seeks necessary services, and requests and manages bank loans. After harvest, member farms pool their products and market them as a single unit in public and/or private markets.

The General Regulation for CCSs, approved in 1993, states that farm holders and their family members may form a CCS of their own volition, according to their individual economic, political, and social interests. As with the CPAs, the highest authority within a CCS is the General Assembly, which elects a board of directors every five years, and makes management decisions according to its own internal regulations. The board is responsible for planning activities and managing the financial assets of the cooperative, and for reporting back to the General Assembly. All expenses are covered by member contributions, and the General Assembly can establish special funds using set percentages of the gross sale of production. These funds can be used to purchase machinery, agricultural equipment, and other items deemed necessary for collective use.

Legal Definition of Property Rights in Cuba

Common property is recognized and codified in national laws that give legal status to cooperatives and protect their right to the land and other productive assets, as contributed by their members. This right is explicitly guaranteed in the current Constitution of the Republic:

- The state recognizes the property of small farm holders as land legally belonging to them as well as the remaining goods and property necessary for its exploitation, as established by law.
- Leasing, agricultural partnerships, land collateral loans, or any other acts implying obligation or transfer of individual rights derived from the land property of small farm holders, are forbidden.
- Small farm holders can incorporate their lands into Agricultural Production Cooperatives; they can sell, exchange, or transfer them for another title to the state, a CPA, or another small farm holder in the cases, forms, and conditions established by law.
- Small farm holders have the right to be associated, in the form and according to the requirements established by law, both for agricultural production and for the acquisition of credits and state services.
- The organization of Agricultural Production Cooperatives is authorized, in the forms established by law.
- Cooperative property is a form of collective property of farmer members.
- The state supports the cooperative production of small farm holders.

(La O Sosa 1997)

Law No. 36 was passed on July 22, 1982, in order to clarify the principals and procedures that define agricultural cooperatives (La O Sosa 1997). This law is currently the primary piece of legislation governing all farmers who choose to work in cooperatives. The first chapter defines the fundamental concepts and precepts for cooperative ownership:

> *Article 1: This law defines the constitutionally recognized right of small farmers to form agricultural cooperatives.*
>
> *Article 2: Cooperatives protected by this law include CPAs, CCSs, and any other cooperative formed in order to increase production.*

This language, taken directly from the Cuban Constitution, explicitly defends the rights of Cuban farmers to own land individually or collectively and to associate for production, credit, and services. Thus it should be clear that a private sector exists within Cuban agriculture that is not a part of the state or of public structures. This does not mean that the government has no role in the private sector, however. Law No. 36 also affirms that the state is responsible for providing economic and technical assistance and qualified extensionists to increase production. However, the state only provides services to the private sector, it is not involved in the management of production or in the allocation of resources.

The State Role in Cooperative Agriculture

In 1993, a major transformation of state agriculture was begun that significantly altered the state's role in cooperative agriculture. Until that time, state farms had provided extensive technical assistance, services, supplies, and extension agents to CCSs and to individual farmers. However, the state farms were slowly broken up into new structures—the UBPCs—that lacked the means to assume the same supportive role, because they also became independent, worker-run cooperatives.

The state increased its efforts to organize CPAs, a process begun in 1977, because the CPAs have a solid and unified structure that works well in the countryside. Human and material resources once available to CCSs were redirected to fund the push for CPAs, gradually diminishing the strength and number of CCSs.

In 1995 the National Association of Small Farmers (ANAP) began a major effort to revitalize the CCSs, which account for more than half of ANAP members. This program provides technical and administrative training to the cooperatives in order to improve management skills and train extensionists to replace the technical services once provided by the state. In coordination with other support measures, the program has been able to strengthen and expand the CCS system. CCSs have increased their contribution to national food supply, and have been able to reach out to the many new, independent farmers who obtained land during the breakup of state farms.

Relationship of CPAs and CCSs to the National Food System

Cooperatives play an important part in the economic and productive capacity of the Cuban countryside. These farmers sell their products through both the state and the farmers' markets. The state buys many of their primary crops, shown in Table 4, either for export or for distribution through the ration system. Not included in these totals is the surplus sold in farmers' markets throughout the country, where prices are set by supply and demand. The latter includes primary crops as well as additional products such as mutton, rabbit, poultry, eggs, and goat's milk. No statistics are currently available to estimate the amount of cooperative-produced foodstuffs that are consumed on-farm, or within local communities. However, the quantity must be significant, since the number of cooperative members and their families exceeds one million people (Álvarez 1999).

Table 4: Contribution to the state sector of Cuba's primary cash crops by CPAs and CCSs

PRODUCT	%	PRODUCT	%
Sugarcane	18	Fruits	68
Tobacco	85	Milk	30
Coffee	47	Honey	51
Cocoa	60	Beeswax	49
Roots and tubers	33	Propolis	59
Vegetables	50	Fish	50
Corn	67	Pork	37
Beans	81	Beef	40

(MINAG 1998)

The contracts that cooperatives have with state buyers are worked out through a complex system that determines the amount and price of products sold based on carefully planned negotiations that include financing, credit, crop insurance, social security, and other state-to-farm relationships. Both individual and cooperative producers, like state farms and factories or unions, participate in formulating the National Economic and Social Development Plan; cooperatives base their annual decisions about planting, sales, and inputs in part on the technical and economic goals for their sector.

All business relations between private producers, or with the state, are governed by contracts that delineate the specific obligations of each party to each other and to the completion of the Plan. There are many different types of contracts within the agricultural sector, including special contracts for the purchase and sale of agricultural goods, produce, supplies, services, transportation, construction, and insurance. The special contracts are an agreement between a producer-seller and a buyer; the producer-sellers in the private sector are the CPAs, the CCSs, and individual farmers, and the buyers are the state enterprises or state organizations in charge of processing and marketing agricultural products. Decree-Law 15/78 of the Council of Ministers covers basic standards for financial contracts in the context of the Plan.

Contracts are also drawn up to provide loans to cooperatives and farmers through the national banking system. Banks charge interest at a rate that varies according to the purpose of the loan. Interest rates on production loans, which cover the current expenses of each crop cycle, and capital improvement loans, vary from four to six percent. Farmers can also request loans at a lower rate of two to three percent to build or repair their homes, roads, and personal infrastructure. The guidelines of the National Bank of Cuba regulate the approval, management, and collection of loans. Collateral is always derived from production. Under no circumstance can land or equipment be subject to foreclosure if a farmer or cooperative is unable to pay off their debts at the end of the loan period.

CCSs can negotiate loans on behalf of their members. They can request standard loans, as well as loans to finance activities or purchases that will benefit cooperative members. CCS loans are repaid from the income generated by the collective sale of the cooperative members' produce.

Agricultural cooperatives and individual farmers are subject to the national tax system. However, the taxation structure differs for agricultural producers in that it offers numerous tax breaks and exemptions as production incentives. Cooperatives must also pay employee taxes for any hired administrative or manual labor, and CPAs are required to pay into a social security fund for their members.

Both state and non-state farmers in Cuba enjoy the protection of a crop insurance system that insures their crops, harvest, and animals against damages or occasional losses due to natural disasters and other unforeseen occurrences.

Determination of Prices

The Ministry of Finances and Prices regulates the price structure for agricultural products within Cuba. The Ministry sets new prices, or adjusts existing prices, by gathering data and opinions from all parties involved. Producers can be directly involved, or participate through political organizations that coordinate with the relevant ministries and other agencies to set prices. Price determination always takes into account production costs in order to guarantee a fair profit for farmers.

Prices vary according to the quantity and quality of the produce. There are special incentives for the production of key products such as export crops or basic foodstuffs. Over the last few years the government has allowed certain regions within Cuba to set minimum prices for fresh produce in order to protect local markets. These measures tend to increase local food production and reduce the losses incurred during the long-distance transportation of fruits and vegetables.

Outside of the state agricultural system, farmers can sell their products in the farmers' markets that were reopened in 1994. Any produce not under contract, or any production that exceeds contracted amounts, can be sold at the markets at prices set by supply and demand. The additional profits garnered through the markets have acted as a strong incentive for producers to exceed their contracted harvests, thus providing more food to the population.

ANAP's Farmer-to-Farmer Training Techniques

Cuban farmers tend to be well educated, with both modern and traditional knowledge. For four decades the Cuban Revolution has been focused on rural development, providing free classes and workshops on both general and agricultural subjects.

ANAP, founded in 1961, is an organization that represents the cooperatives and individual farmers that make up the non-state sector. Today its primary goal is to encourage and develop the use of agroecological farming techniques throughout the Cuban countryside. Some of its activities include:

- nationwide training programs to build capacity among small farmers, cooperative members, grassroots organizations, and ANAP leaders
- farmer-to-farmer training programs where farmers teach each other about their experiences with sustainable agriculture through direct participation and communication
- reorientation of the National Training Center's education and training curriculum in order to emphasize agroecological knowledge
- collaboration with international donors and nongovernmental organizations (NGOs) to promote sustainable techniques
- farmer, extensionist, and researcher participation in regional and national networks to discuss topics related to food security and sustainable development

ANAP combines traditional knowledge and practices with new technologies in a participatory effort that enables farmers to educate each other. The organization is broad-based and horizontal in structure. Although its headquarters are located in the Niceto Pérez National Training Center, the majority of ANAP's activities are decentralized through provincial and municipal offices. Planning meetings and programs are held at regional locations appropriate to the topics discussed, be it at ANAP facilities, local Ministry of Agriculture (MINAG) offices, or on-site at farms or cooperatives. This ensures that meetings will be comfortable, accessible, and inclusive. This inclusive model of communication has had great success in rural Cuba.

Campesina *explaining how to tend transplants at a coffee nursery.*

Grassroots networking and extension dates back to the early years of the Revolution, when the lack of qualified personnel in rural areas appeared to be a powerful obstacle to development. The few available technicians taught agricultural and veterinary techniques to farmers, who would then become trainers themselves. This technique, arising from necessity, was tremendously successful. In 1961, similar methods were used to combat rural illiteracy, which at that time was approximately 40 percent. Because of these early campaigns, farmers are well prepared for grassroots education.

Farmer-to-farmer programs are particularly useful when promoting ideas or techniques that contradict conventional wisdom and customs. In order to change deeply rooted habits, the teacher must establish a high level of confidence and credibility with the student. Through farmer-to-farmer contacts, ANAP has been able to maintain a strong relationship with its members. Thus, it has been very successful in disseminating teachings from scientific and technical institutions through its national structure, allowing the information to reach even farmers in remote areas.

Some of ANAP's training is conducted via the media. Nationwide, ANAP hosts regular programs on more than fifty radio stations, most of them community based. ANAP has created television shows specifically for farmers that reflect their lifestyle and cultural heritage and provide technical information and training. ANAP's magazine reports on the recent agricultural news and scientific knowledge, including the theories and practices of agroecology. Promotional materials provide information on specific pests and diseases, biological pest controls, agroecological techniques, natural food preservation, and other topics.

However, the most efficient method of transmitting information and building consciousness in rural communities is still direct, personal communication. During the Special Period the economic crisis limited access to printing and publication materials. Therefore, the farmer-to-farmer training schools have remained the crux of all outreach efforts. ANAP's successes over fifteen years of communication and training for rural activism earned them UNESCO's International Communications Development Program Award in 1989. (UNESCO 1989a; UNESCO 1989b).

Extension work, called the Continuous Teaching and Education Program, is part of the Small Farmer Technical Activism Program. Participating farmers attend an intensive training course at the Niceto Pérez Center during which the farmers themselves prepare the materials (both written and audio-visual) for extension work in their region. The participants then spend some time in the provincial offices, collaborating with extension agents, technical specialists, and ANAP leaders in order to develop a provincial plan. The entire team of extension agents and specialists is then dispatched throughout the area to organize meetings, teach workshops, and coordinate with local community activists. Once regional training has been completed, farmers from various regions come together in larger exchanges in order to compare experiences and discuss their work. Finally, team members evaluate the impact of their extension work in meetings with participating farmers, and a new cycle begins as another group of farmers travels to the national center for training courses.

These extension methods, commonly called "farmer-to-farmer" exchanges, identify and emphasize many traditional farming practices that are both productive and conserve and rehabilitate farming ecosystems. Cuba's ability to survive both natural disasters and economic crises is derived from the cultural strength of its rural population. Farmers are closely connected to their land, able to observe and adapt to changing conditions. Throughout the difficulties and shortages of the 1990s, farmers have had a sizeable impact by guaranteeing the food supply for Cuba's population. They have been able to maintain, and even increase food production, in many cases using exemplary methods of sustainable agriculture.

The Role of ANAP in Promoting Sustainable Agriculture

Since the collapse of the Soviet Bloc, agricultural production in Cuba has been proceeding under extremely difficult financial circumstances. The powerful effects of the US trade embargo have been intensified, restricting the supply of conventional imported agricultural inputs like oil, machinery, and agrochemicals. Cuba has also been the victim of intermittent attacks by introduced biological warfare agents designed to damage both animal and crop production. These factors, combined with Cuba's geographic predilection for natural disasters (cyclones, hurricanes, and tropical storms), have impinged on the ability of farmers to maintain production levels.

Farmers in Cuba have turned to sustainable agriculture as a means of survival during these difficult circumstances. The transition to sustainable techniques has been easier for Cuban farmers than in other countries because of the security bestowed by the Cuban government: land rights, access to and ownership of equipment, availability of credit, markets, insurance, and free, quality health care and education. Sustainable technology is difficult without sustainable economic and social structures. Cuban farmers are highly organized through the formation of cooperatives with real social and economic power, and the presence of national organizations such as ANAP that can represent the interests of individual farmers at the state level.

ANAP members added sustainability as one of the Farmer-to-Farmer Extension Program's official goals at the VII International Meeting in November 1996, in the Guira de Melena municipality of Havana Province. Representatives attended the meeting from Mexico, Central America, and the Caribbean; many had been working in solidarity with Cuba since the early 1990s to promote agricultural exchange and friendship throughout the region (ANAP 1997a).

ANAP defined its commitment to sustainability and agroecological agricultural through three basic goals:

- to restore and promote the practices of small farmers through direct farmer-to-farmer exchanges of sustainable agricultural techniques
- to support horizontal technology transfers through participatory methods that encourage the use of appropriate sustainable technologies
- to conduct the research necessary to carry out successful agroecological extension, public education, and appropriate technology transfers

In a joint project with Bread for the World (Germany) and the Council of Churches of Cuba's Department of Project Coordination and Assistance (DECAP), ANAP initiated its agroecology program in Villa Clara province. ANAP identified more than 200 small farmers who were already experienced with and practicing agroecological techniques, half of whom began to work as extension agents through the farmer-to-farmer program. In 2000 similar programs began in the Cienfuegos and Sancti Spíritus provinces, with expansion planned to the rest of the country. The province of Holguín has already begun agroecological training driven by local, farmer-sponsored initiatives.

The Role of Other Institutions in Promoting Sustainable Agriculture

Cooperatives have played an important role in educating tens of thousands of farmers about agroecology. Extensionists give workshops and classes at General Assembly meetings of CPAs and CCSs, utilizing existing organizational structures in order to facilitate the educational process. Given that there are currently over 3,700 cooperatives with monthly meetings, this structure has been a highly effective way to teach farmers about the values of sustainable agriculture.

ANAP has also worked with the Center for the Study of Sustainable Agriculture (CEAS), a part of the Agrarian University of Havana (UNAH). Since 1998, the National Training Center and CEAS have been offering the Agroecology and Sustainable Rural Development Chair at the UNAH for professors committed to sustainable agriculture. The impact of this program is multiplied through the national network of research institutions, and the farms and cooperatives that provide practical and demonstrational points of reference.

Every year, CEAS trains thousands of students, administrators, and farmers in modern agroecological principles and technologies. In the first trimester of 1999, 36 ANAP employees and cooperative members received their diplomas. ANAP is currently working to make this type of certified, technical training available throughout the country by means of the decentralized training system used in their own extension work.

Within Cuba, ANAP has coordinated with the following promoters of sustainable agriculture: the Organic Farming Group (GAO) of the Cuban Association of Agricultural and Forestry Technicians (ACTAF), DECAP of the Council of Churches of Cuba, the Center for the Study of Sustainable Agriculture (CEAS), the National Institute for Fundamental Research on Tropical Agriculture (INIFAT), the National Center for Plant Protection (CSNV), and the National Plant Protection Institute (INISAV), Soils Research Institute (IIS), Liliana Dimitrova Horticulture Research Center (IIHLD), and Agricultural Mechanization Research Institute (IIMA), among others.

International Cooperation

In 1993, ANAP began to make contact with international donors, as its educational and training programs required financial support, particularly for printing and publications, audio-visuals aids, and transportation. Because of the economic situation, Cuba's exports had shrunk, and the hard currency necessary to purchase these services was only available from foreign sources.

Relationships with foreign NGOs have provided much-needed alliances and solidarity, made difficult by the increasing stringency of the US blockade. The first alliances were made with Nicaragua, via a Norwegian NGO, Norway Popular Assistance. ANAP then established relationships with the "Southern Group," composed of the Association for Cooperation with the South (ACSUR, Spain), Oxfam Belgium, and *Tierra de Hombres.* Offers of assistance and solidarity have grown since then and, by 1999, ANAP was collaborating with fifty NGOs from more than 20 countries.

In 1996, ANAP began working on North-South-South cooperation, or "trilateral aid." They developed a project to establish agricultural cooperatives among small individual farmers in the municipality of San Antonio de los Baños in Havana Province. The fiscal sponsor of the project is a Belgian NGO called National Center for Development (NCOS). NCOS funds an exchange between ANAP and the Uruguayan Cooperative Center (UCC), which assists with technical advising, project implementation, and other aspects of the project (Álvarez 1997). Many NGOs in Latin America have also sought funds from northern NGOs in order to provide scholarships for their farmers to attend Cuban schools and training programs and to participate in meetings and exchanges in Cuba.

These collaborations have served to strengthen the sustainable agriculture movement by building communication and solidarity across borders. It has also allowed Cuba access to new resources, both material and intellectual. Different NGOs have collaborated and continue to support ANAP in this task of promoting a viable and sustainable agriculture among small farmers. The following donors have strong collaborative relationships with ANAP: the Spanish organizations *Fundescoop,* Trade Union Institute for Cooperation in Development (ISCOD), *Largo Caballero* Foundation, Center for Rural Studies of the Polytechnic Institute of Valencia, and *Entre Pueblos;* Oxfam Belgium; Civil Volunteer Group of Italy; and the German NGO Bread for the World (Simón 1999). ANAP is also a member of *Via Campesina* and a Regional Coordinator of the Farmer-to-Farmer Program of the International Network for Agriculture and Democracy (RIAD).

Conclusion

Despite the difficult circumstances of the Special Period, Cuban agriculture has undergone a positive transformation toward sustainability. An increasing number of farmers are abandoning the conventional production model that was based on excessive use of agrochemicals, wasteful use of resources, pollution of water and soil, destruction of forests and ecosystems, poor soil management, and many other misdeeds that have left our planet on the verge of ecological catastrophe.

In Cuba, farmers, extensionists, and researchers are collaborating to promote and apply a combination of traditional agriculture and modern scientific and technical knowledge. These alternative technologies are spreading throughout the Cuban countryside despite shortages in supplies and a deeply rooted predilection to use agrochemicals.

Cuba continues to develop and implement new technologies. Research centers have developed improved techniques for soil management such as crop rotation and crop/animal integration, application of animal manures, composting, green manures, and the use of animal traction instead of heavy machinery. Biological pesticides and fertilizers are becoming increasingly mainstream as Cubans recognize the need to reduce toxic chemicals. Cuba is planting neem trees (*Azadirachta indica*) and using the potent botanical pesticides it produces to combat insect problems. The first two neem production plants were built on CPAs in the provinces of Matanzas and Havana (ANAP 1997b). The Centers for the Production of Entomophages and Entomopathogens (CREEs) are producing at capacity, and still the demand is rising for predatory insects, bacteria, and fungi, in both rural and urban areas (ANAP 1997c). Recently, Cubans have begun to adopt alternative energy sources such as RAM pumps that use gravity to pump water out of wells, and the biodigestors that create cooking gas and potent biofertilizers out of pig manure.

These new technologies have allowed small and medium farms to become a significant source of Cuba's food supply while protecting the surrounding environment. Farmers have gained a greater understanding of the need to protect natural resources and to practice a healthier, more harmonious, and balanced agriculture.

ANAP, as the principal organizational structure for small farmers, has committed itself to sustainable rural development, and created a clear strategy to strengthen farmer's cooperatives. Their programs are designed to allow farmers to increase production, thereby increasing food security among the Cuban population, using sustainable techniques. These farmers then become teachers and examples of modern sustainable agriculture.

References

Álvarez, Mavis D. 1997. *El futuro de la cooperación internacional en el marco institucional de la ANAP.* Informe al Buró Nacional de la ANAP.

Álvarez, Mavis D. 1999. Estructuras de Producción de las CPA y CCS. *Documento interno de la ANAP Nacional.*

ANAP. 1997a. Promoción productiva agroecológica de *campesino a campesino.* Proyecto de Cooperación al Desarrollo ANAP–Pan para el Mundo, CUB-9710-004. October. Dirección de Cooperación ANAP.

ANAP. 1997b. Utilización del potencial agroecológico del árbol del Nim. Construcción de una planta procesadora para frutos y semillas del Nim en la CPA Cubano-Búlgara. Proyecto de Cooperación al Desarrollo ANAP–Sodepaz-GTZ. November. Dirección de Cooperación ANAP.

ANAP. 1997c. Extensión de las técnicas de empleo de los productos biológicos en el control de plagas en la agricultura. Proyecto de Cooperación al Desarrollo ANAP–Oxfam Solidaridad de Bélgica. Dirección de Cooperación de la ANAP.

ANAP. 1998. Dirección Nacional de la ANAP. Informes Estadísticos Anuales.

La O Sosa, M. 1997. *Compendio de Legislación Agraria y Documentos de Interés para el Trabajo de las Cooperativas de Producción Agropecuaria y de Créditos y Servicios.* Prensa Latina S.A. Havana: World Data Research Center.

MINAG. 1998. Ministerio de la Agricultura. Boletines estadísticos sobre la producción del sector cooperativo y campesino. Havana: MINAG.

Simón, F. 1999. Cooperación ANAP-ONG. *Revista ANAP* 38:3.

UNESCO. 1989a. Recibe la ANAP premio PIDC de comunicación rural. *Boletín de la Comisión Nacional Cubana de la UNESCO.* January–June, pp. 18–19.

UNESCO. 1989b. Sources. *Planéte.* Prix PIDC pour la communication rurale. No. 3 April, pg. 21.

CHAPTER SIX

Agroecological Education and Training

Luis García, Center for the Study of Sustainable Agriculture, Agrarian University of Havana

"Education is critical for promoting sustainable development and improving the capacity of the people to address environment and development issues."

—AGENDA 21, CHAPTER 36.2

Around the world the modern industrial model of agriculture is in crisis. As basic natural resources for the production of food and fiber are eroded and contaminated, the vision of ever increasing yields through industrialization of production is being challenged. As a result of this crisis, a new model of agricultural development is gaining favor—*agroecology*. This new paradigm views the farm as an ecosystem, and blends the technological advances of modern science with the time-tested and common sense knowledge of traditional farming practices. As this standard gains increasing favor through grassroots and policy efforts, the need for trained professionals and technical and managerial staff is increasingly felt (Vandermeer et al. 1993, FAO-ORLAC 1995; Rosset 1997). In this sense Cuba is no different from most of the rest of the world. However, the need has been felt more urgently in Cuba in the last decade due to the crisis-induced scarcity of external inputs for agriculture.

The conventional model that universities have used to train agricultural professionals and extensionists since the 1950s and 1960s is based on the industrialization of agriculture. This model depends fundamentally on high response seed varieties, which need heavy doses of synthetic fertilizers, chemical pesticides, heavy machinery, and often irrigation—the basic elements of the so-called "Green Revolution." Although in the last 30 years Green

Revolution technology has allowed important gains in per capita world food production, rural poverty has also increased, as the technologies were developed for large agricultural enterprises, thereby leaving small farmers at a disadvantage. It has also accentuated an already unequal global distribution of food, favoring those countries and sectors with greater economic resources.

Beyond the well-documented destruction of the natural resource base necessary for future production, the social and economic impacts of this model are the greatest threats to its long-term sustainability. Looking at production costs alone, this model is largely unsustainable, as evidenced by the sizeable subsidies required to keep it functioning. In 1992, in developed countries alone, industrial agriculture required the enormous subsidy of $356 billion dollars to keep it afloat (FAO-ORLAC 1996). Seen in this light, the need to transform world agriculture to an economically and ecologically sustainable model is undeniable. Such a new model must meet the requirements of both present and future generations, in ways that are socially just and culturally appropriate (Altieri 1995).

In Cuba, the transition to a new model is well underway, and has generated a huge demand for agroecological education programs, to produce agricultural professionals with the theoretical, technical, and practical training required to make the new model successful. The essential areas of training (and re-training) necessary for this massive transition reflect the sustainable agricultural model being implemented on a national scale. This new model prioritizes the following areas: organic fertilization, soil conservation and recovery, the use of draft animals, alternative energy sources, integrated pest management, crop rotations and intercropping, on-farm integration of crops and livestock, alternative veterinary medicine, urban agriculture, cooperative land use, the adaptation of cropping systems to local conditions, a reduction in the scale of farming, with increased community participation, and alternative methods of research and education (Garcia, Pérez, and Marrero 1997). This final element, research and education, plays a decisive role in all of the others.

The gradual introduction of this sustainable model, adapted to local conditions, has already reduced some negative environmental impacts and achieved modest but sustained increases in food production over the last two to three years. Given the fundamental importance of education for the sustainable development of society in general, and of agriculture in particular, in this chapter I analyze some key experiences in training and education for sustainable agriculture in Cuba.

Potential for a National Transition to Sustainable Agriculture

Canadian-American professor Dr. Patricia Lane has declared that the new Cuban development model might just allow Cuba to become one of the first sustainable societies of the twenty-first century (Lane 1997). In her analysis she highlights scientific and educational development as one of the basic pillars of sustainability for this Cuban new model. The model is well rooted in the history of Cuban nationalism, and particularly on the ideological and pedagogical legacy of José Martí. The Cuban economic crisis of the 1990s, caused by the US blockade, provided a unique opportunity for the collective creation of alternative development approaches, based on scientific knowledge in search of practical solutions, to address many of the environmental problems that humanity faces today (CIDEA 1997).

As Fidel Castro has argued, without the work carried out since 1959 in educating the Cuban population, it would have been impossible to survive the "Special Period" and continue developing (Castro 1997). During the initial years of the Special Period, there was an overall 30 percent decrease in imports, which led to a substantial reduction of food availability for the Cuban people. In the worst moments of this crisis, average food consumption dropped to 60 percent of basic nutritional requirements, and for some food groups, such as fats, it was only 35 percent of nationally recognized standards (República de Cuba 1994). Cuba survived this crisis only because of the high level of organization and education that our society has achieved.

For more than three decades, Cuba worked to reorient its economic trajectory and face the challenges of the future. Today we have an acceptable productive infrastructure, a highly qualified work force, and considerable scientific and technical potential. One in eight Cuban workers has at least a vocational degree, one in 15 has a university degree, and one in 900 has a Ph.D., all very high numbers for the Third World (República de Cuba 1994). This development of "human capital" only became a top priority after 1959, when illiteracy still affected 24 percent of the population. By the early 1960s, illiteracy was practically eradicated. By the 1980s, the average educational level of the working population had increased from second to ninth grade. In spite of these tremendous advances, as it stood in the early 1990s, the overall progress in education was still not sufficient to address the specific needs of a national transformation to a more sustainable agricultural system.

It is widely acknowledged that the transition to sustainable agriculture requires substituting capital-intensive technologies with knowledge-intensive ones, thus requiring a more educated and skilled labor force (CLADES-FAO 1991). The key factors for a successful conversion to sustainable agriculture in Cuba are the training and education of farmers in specific production techniques, strong social organization in the farming sector, and the development and application of new appropriate technologies. Among these, training and education play the primary role, though all elements are extremely important (FAO-ORLAC 1995).

With the onset of the Special Period, agricultural and livestock professionals suddenly found themselves technically unprepared for the Cuban agriculture of the 1990s. Cuban agronomists were highly trained and skilled, but their skills and training were inappropriate for low-input production systems. Their training had been completely based on Green Revolution technology, and without chemical inputs, machinery, and petroleum, many did not know where to begin.

In response to this problem, the agricultural universities, led by the oldest among them, the Agrarian University of Havana (UNAH), initiated courses to update their graduates and other professionals with agroecological training as quickly as possible. Soon after the Center for the Study of Sustainable Agriculture (CEAS) was created to provide a focal point dedicated to supporting the new research and education needs of the agricultural scientific community. These efforts resulted in important modifications to the undergraduate curricula. Over the course of the 1990s, agroecological theory and practice were incorporated into undergraduate programs in every agricultural field, and existing master's and doctoral specialties were loaded with agroecological content. Additionally, there are now general agroecology master's and doctorate programs at the UNAH and at several other leading provincial universities.

Today regional training programs are being defined according to local and regional needs. These programs need to be evaluated in terms of the concrete objectives of productive projects, at the farm, province, and national levels, rather than by traditional indices such as numbers of students or course hours. One estimate suggests that the use of known and existing low-input techniques, which can be readily incorporated through training, would make it feasible for Latin America to boost agricultural production by as much as 40 percent (UNESCO-OREALC and FAO-ORLAC 1988). Cuba is trying to reach that goal, while developing new and innovative technologies to surpass it.

Needs Assessment for Agroecological Education and Training

The first steps in developing a new learning process are to determine who needs training and to define the kinds of training that different populations need. In the agricultural sector, farmers, technicians, extensionists, professionals, and university students are usually the target populations (FAO 1993). However, it should be noted that the theory and practice of agroecology should not be artificially separated from the larger questions and audiences associated with the development of sustainable systems in other sectors and in society at large. The objectives of each coincide, and thus they must interact with one another. It is impossible to attain sustainable development of society without a sustainable agricultural sector and the safe food system it produces, and vice versa.

Agroecology training workshop, Camagüey province.

In this light it is evident that training of traditional target groups is not enough. In Cuba we have seen that it is critical to offer new training to the directors of state agricultural enterprises, policymakers and production managers, to the members of agricultural cooperatives, and to the general public in rural areas (estimated at more than two million people). If we also take into account the growing role of urban agriculture in Cuba, then we must also include the eight million urban dwellers. Here the public must not only be educated in their role as consumers, but also with regard to their direct role in urban gardening and farming.

Thus a comprehensive quantitative determination of training needs would include almost all of the 11 million Cubans. Of course the specific agroecological content and its technical level differ considerably with respect to each population, as do the objectives, methods, and means used to reach them. Clearly this is a massive undertaking, so agricultural professionals and workers must receive the highest priority, while not forgoing an educational effort directed at the general public. Making a priority list, from both the private and public sectors, yields a total of more than 750,000 people (see Table 1).

Table 1. Priority human resources in Cuban agriculture, 1996

CATEGORY	NUMBER
Professionals (university degree or higher)	19,390
Trained technicians	56,505
Credit and Service Cooperative (CCS) members	121,070
Agricultural Production Cooperative (CPA) members	34,898
Basic Unit of Cooperative Production (UBPC) members	115,522
Managers and workers of state enterprises	414,220
Others	14,796
Total	768,401

Although it is difficult to give exact numbers, based on course enrollments at various institutions and associations, we estimate that approximately 100,000 people attend some form of agroecology training each year. Some 1,000 of them attend a yearlong theoretical and practical university course or a graduate level course.

From a centralized viewpoint it is easy to determine who needs to receive training in a given field just by examining the objectives of the different organizations, institutions, associations, and other agencies in that sector. But it is much more important to find those people who are actively seeking out training based on a strongly felt need.

An important point in this sense is that voluntary, self-initiated applications to enroll in agroecological programs are significantly higher than in other fields of agriculture. For example, when the first certificate course for professional training in agroecology and sustainable agriculture was offered in 1995 for professionals in 11 of the 14 provinces, enrollment was 10 to 20 times higher than in other fields of study. This undoubtedly demonstrates a "felt need" or social necessity and the relevance and importance of this information.

The National Strategy for Cuban Environmental Education singled out the training of professors, decision-makers, and journalists as a high priority (CIDEA 1997). This is in line with the needs of the agricultural sector where intensive work on training intermediate and higher-level professionals has already begun; though some education has also been provided to the other two groups. In the case of *campesino* leaders, several thousand have received agroecology training at the National School of the National Association of Small Farmers (ANAP), where the curriculum was developed in collaboration with CEAS. The training at the National School is a key element in the agroecological movement, due to the influence of these farm leaders in their respective cooperatives.

Many groups have recognized the need for agroecological education of children, because it is clearly better to educate correctly the first time than to try to reeducate later. A number of institutions and NGOs have been working in collaboration with the Ministry of Education to introduce agroecological concepts to elementary school children. This is done through both formal and informal methods such as extra-curricular activities and clubs (Pérez 1997).

For decades Cuban professionals, technicians, and farmers were educated in high-input, industrial agriculture. As a consequence, much of their basic knowledge is useful, though deeper knowledge and skills for low-input agroecological agriculture are required. Through learning to use technologies and approaches that differ from the industrial model, their most basic concepts of what agriculture is and how to do it, are challenged. In general, to be able to do organic or low input sustainable agriculture, farmers and professionals need conceptual reorientation so that they can view the agricultural landscape as an ecosystem rather than simply as a production unit. They need training that focuses on ecological concepts such as the relationships between structure and function in different agroecosystems (Altieri 1995). A new educational paradigm is needed which not only creates such an agroecological consciousness, but which is also capable of changing ingrained and unsustainable behaviors.

In Cuba, the National Subsystem of Agricultural Education has gradually incorporated the sustainability dimension into its curriculum. The Organic Farming Group (formerly ACAO, now part of the Cuban Association of Agricultural and Forestry Technicians, ACTAF) has played a strong role in energizing this effort. They have done this by collaborating with the National Education System, the Ministry of Higher Education, the Ministry of Agriculture, ANAP, and specific universities and research centers (see Table 2) to develop and provide quality educational opportunities in this field.

Table 2. Key actors in agroecological training in Cuba

Universities
Vocational high schools (IPAs)
Research centers
National Association of Small Farmers (ANAP)
Training centers of the Ministries of Agriculture and Sugar
Agroecological Lighthouses (UNDP-SANE Program)
Farmers (and their cooperatives), researchers, and professors
National and international nongovernmental organizations (NGOs)
Council of Churches of Cuba
Mass media

While all of these actors have carried out successful projects at different geographical scales, a systematic approach to agroecological education is really needed at the farm, provincial, and national levels. Early experiences in the organization and coordination of these activities on all three scales have clearly demonstrated that this is the most successful way to address the problem.

Farmer and Farm Manager Training

In addition to the thousands of people trained by the universities, there are tens of thousands trained by the Ministry of Agriculture and ANAP. They use courses, meetings, workshops, field days, talks, and experiential exchanges. Of particular note are the courses given by the Institute of Veterinary Medicine on traditional medicine and acupuncture for livestock, as well as the conferences and courses developed at the National School of ANAP, that have had outstanding results. The use of farmer-to-farmer methodology has engendered the participation of more than 600 farmers in the City of Havana, and achieved an amazing transformation in the Santa Fe municipality.

Regional organization and collaboration with the agroecological training effort is also necessary. The cooperation of different institutions combined with the proactive role taken by the universities (who are responsible for professional and technical training in each province) has greatly expedited the introduction of agroecological learning and educational plans. For example, the province of Havana is a leader in the organization and planning of agroecological training and now boasts a network of organizations and institutions led by CEAS in 9 of the 19 municipalities. Similar advances have been made in the eastern province of Guantánamo where special consideration is given to farming on steep slopes. In the far western province of Pinar del Río, agroecology has become an important part of a more extensive environmental conservation effort.

At the community level, the Agroecological Lighthouse Program of UNDP's Sustainable Agriculture Networking and Extension (SANE) Program has generated exemplary results working with a network of Agricultural Production Cooperatives (CPAs). This project provides specialized agroecological training programs to farm leaders. The most outstanding students join the master's program offered regionally. At the same time, training is offered to the general coop membership and to farm workers. This training has contributed significantly to the overall success achieved by this program. Because of the wide range of participants who need training and education, the SANE staff organizes programs at different levels and quite often in nonconventional ways.

In addition to formal education through the National Education and Training System and the regional university and NGO programs, nonformal education efforts are equally important. Other organizations play a significant role in disseminating information, such as in the extracurricular activities organized by both community and national institutions. These might be social activities, scientific and cultural events, or demonstration production operations. News coverage of these events and activities has also played a critical role in reinforcing the importance of these efforts to the participants and in educating the general public through the mass media (CIDEA 1997).

Agroecology at Vocational High Schools (IPAs)

In Cuba the rural vocational high schools, called Agricultural Polytechnic Institutes (IPAs), are the first line of agricultural education and training. It is here that future farmers and agronomists get their first formal exposure to the science and technology of agriculture. It is therefore critical that these schools have strong programs preparing the country's future agroecologists. There are agricultural programs offered at 143 polytechnic high schools throughout the country, of which 111 are in agronomy, 17 in livestock production, and 15 in mechanization. This represents a major expansion during the Special Period, as there were only 55 in 1990. There are approximately 41,300 students enrolled in these programs who upon graduation will be certified for specialized jobs in agriculture and livestock production, or will go on to university studies (MINED 1996).

All these high schools have agroecological production and experimental areas to give students practical experience. This provides for the development of graduates who are innovative and adaptive, and who can apply new technologies in ways that balance maximum productivity with environmental agroecosystem stability—this being the essence of alternative agriculture (Santa Cruz and Mayarí 1997). At each school students take courses and work daily in the fields. Each school produces food for the students, on their production plots, and there are also research plots for student projects. Generally they are full-time live-in schools. Because they are producing their own food, many of the schools have highly diversified farms with a high degree of crop-livestock integration. At the same time each school specializes in the important crops of the region, so that each may have specialties and facilities not found at other schools. Many of the schools produce their own biological pest control agents, and a few run Centers for Production of Entomophages and Entomopathogens (CREEs) which sell to local farmers, generating revenue for school financing.

From the 1991–1992 to the 1996–1997 school years, progress in the application of alternative techniques in agriculture was quite impressive. Particular highlights were the establishment of biological control lines; the production and use of biofertilizers; different techniques for the production of organic fertilizers (hot composting; worm composting or vermiculture); the use of biointensive methods for vegetable production; the increased use of local community resources (human, natural, and waste); as well as recovering non-productive marginal areas (see Table 3).

Table 3. Sustainable agriculture programs at vocational high schools, 1991 to 1997

SCHOOL YEAR	NO. OF SCHOOLS	ORGANIC FERTILIZERS	VERMICULTURE	COMPOST	BIOLOGICAL CONTROL (lab/field)	MEDICINAL PLANTS
1992–1993	154	2	42	43	11/30	2
1993–1994	164	2	100	105	20/25	10
1994–1995	163	2	105	107	30/35	14
1995–1996	150	3	107	110	38/45	76
1996–1997	143	3	112	112	50/52	102

Note: The 102 schools producing medicinal plants averaged 30 plant species per school and used them directly for student health care in accordance with practices established by the Ministry of Health (MINSAP).

(After Santa Cruz and Mayarí 1997)

The high school teachers are trained by a variety of means, including joining forces with local farms and production centers. In the latter case, they can both benefit from regional training programs for farmers and farm managers, and also be better able to adjust their teaching to the needs of local production. A new approach to curriculum development has been created to define the objectives, technical knowledge and skills that must be attained by students—a "production-based, techno-pedagogical needs analysis."

The Role of Universities

It is generally accepted that universities can and must play a fundamental role in the transition to sustainable agriculture. Considering the role universities play in Latin American society, their political importance, their research and development potential, and their responsibility in forming new generations of professional personnel, such a transition at national and regional levels is truly unthinkable without them (CLADES-FAO 1991; Sarandón and Hang 1995; García 1997).

Although not yet true of the majority of Latin American universities, in Cuba academia has played a critical role since the beginning of the transformation toward sustainable agriculture (Funes 1997). The agroecological concept has been widely introduced into the curricula of the universities, presented as a system of knowledge, abilities, attitudes, aptitudes, and values, and begins with general educational objectives and includes all specialized course-work and training. The objectives are derived from the practical needs of each sector, and study programs are designed using Cuban methodology (Martínez 1991).

Considered transdisciplinary in focus, agroecology has blended with other disciplines, including social sciences and other subjects, reorienting the way future professionals in diverse fields will view agriculture in society, and integrating other fields more closely with professional agricultural practice. From the beginning, interdisciplinary expert panels, surveys, workshops, and exchanges between farmers, students, agricultural employers, researchers, and professors were used to improve existing programs and to develop the new agroecological curriculum.

At the university level, course curricula have been developed to give a general overview of agriculture as an ecological science, and to train professionals in the specific areas they will need before entering the field. The university curricula and professional training programs overlap considerably in terms of content, but the professional courses assume a high degree of previous knowledge as most of the professionals are already graduates of technical or university programs. Based on needs assessments completed for both levels, the following topic areas are now covered in varying degrees throughout the study programs:

- social and environmental impacts of the industrial agricultural model
- theory, practice, and development of agroecology, organic agriculture, and sustainable agriculture
- the different components of the Cuban model of sustainable agriculture
- structure and function of agroecosystems
- diagnosis and design of sustainable agroecosystems
- methodologies for technical analysis and evaluation of agroecosystems
- methodologies for socioeconomic analysis and evaluation of agroecosystems
- indicators of sustainability
- use and evaluation of biodiversity in agroecosystems
- integration of aquaculture into agroecosystems
- design and evaluation of agricultural projects
- ecological economics
- certification and marketing of organic products
- regional planning and watershed management
- minimum and zero tillage
- bioethics

- soil and water use and conservation
- alternative energy sources
- ecological agrometeorology
- livestock production in sustainable systems
- agroecological soil management and soil fertility
- agroecological pest management and integrated pest management (IPM)
- intercropping and crop rotation
- agroforestry and silvopastoral systems
- sustainable forest management
- crop–livestock integration
- draft animals and animal traction
- traditional and green veterinary medicine and acupuncture
- mechanization and sustainable agriculture

At the post-graduate level (master's and Ph.D.), the following additional topics are covered:

- research methods in sustainable agriculture
- research seminars
- multivariate statistics, experimental design, nonparametric statistics, and production statistics
- ethnoecology and traditional knowledge in agriculture
- general ecology
- environmental impact analysis
- production of biological control agents
- training and education methods for agroecology
- farming systems of Latin America and the world
- rural sociology
- economics and rural development
- epistemological aspects of agroecology

All of the undergraduate, certificate, master's, and doctorate level curricula have now been officially accredited by the corresponding educational agencies. The National Commission on Postgraduate Education and Scientific Qualification has approved all of the programs and course materials for all the university and research centers. Table 4 presents the master's program at the UNAH.

Table 4. Master's Course Curriculum (UNAH)

REQUIRED COURSES	HOURS	CREDITS
General agroecology	45	3
Research and extension methodology in sustainable agriculture	30	2
Statistical analysis and experimental design	60	4
Thesis seminar/Workshop I	20	1
Agrometeorology	30	2
Soils and their agroecological management	60	4
Ecological pest management	60	4
Economics and agricultural development	45	3
Rural sociology	35	2
Certification and marketing of organic products	35	2
Agricultural production project implementation	35	2
Agroforestry and nature conservation	30	2
Crops in tropical systems	45	3
Animal production in sustainable agroecosystems	45	3
Thesis seminar/Workshop II	15	1
ELECTIVES		
Traditional veterinary medicine and acupuncture	45	3
Rotational grazing systems	30	2
Production of biological pest controls	60	4
Mechanization in sustainable agriculture	45	3
Ecology	45	3
Weeds and their ecological control	30	2
Rural development and cooperatives	30	2
Thesis defense	30	–

Note: To graduate each student must also publish at least one scientific paper.

An integrated curriculum has been designed and introduced throughout the country by the CEAS. It includes a reorientation of basic undergraduate agronomy degrees, short courses of 40–60 hours with practical training on specific techniques for professionals in the field, an academic certificate degree on agroecology and sustainable agriculture which includes three modules and corresponding manuals, a master's degree (M.Sc.) and a specialization in the doctor of agricultural sciences (Ph.D.). degree (see Figure 1).

Figure 1. Flow chart of postgraduate level training in agroecology

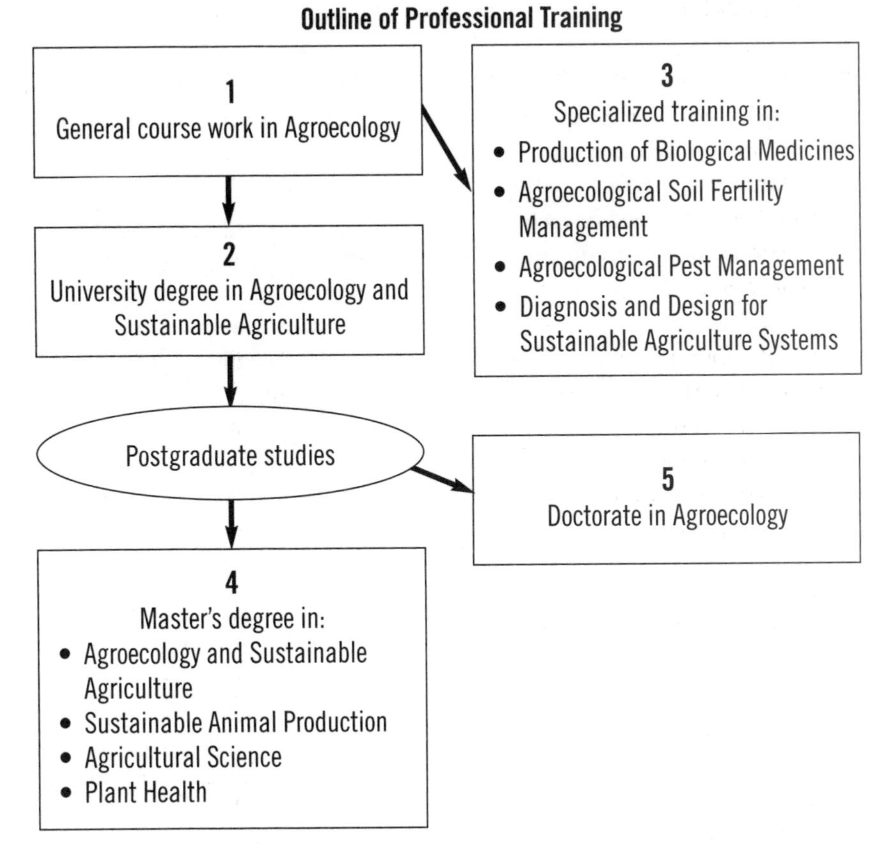

All of these form part of the national university structure, offered in different provinces with students from 11 of the 14 Cuban provinces and the Isle of Youth. Many of the programs are designed for working professionals and offer both full-time and part-time options and/or distance courses. The evaluation of these post-graduate programs by a Spanish university found them to be the best in Cuba's agricultural sector, and the master's program was found to be among the best of any academic field in the country.

As both master's and doctorate programs require original research, the role of these programs in research and development of new agroecological theory and practice is also very important. There are three National Scientific and Technical Programs related to sustainable agriculture that are dedicated to funding and overseeing new research and development projects. The Research Faculty of the UNAH has directed more than half of these projects. This represents a great opportunity for students to get research experience and to create their own thesis projects.

Pedagogy for Agroecological Education

While at first glance the essential differences between agroecological and conventional agricultural education are in the course content, the real contrasts and contradictions are in the basic philosophy of science. Conventional agronomy assumes you can break down any system to its basic production factors, separate and study them independently, and then reassemble your findings such that the total is equal to the sum of its components. Agroecology assumes that the interrelatedness of each production factor to the others is so profound that to study them you must take an interdisciplinary, holistic, and systemic approach. Thus, changing the content has required creating a new educational methodology. This new model has its own basic principles and characteristics consistent with the principals of environmental education. Like the field of environmental education, this new methodology, in accordance with the Cuban Strategy for Environmental Education, must be "active, flexible, and participatory, stimulating creativity and intelligence," where the subject-object relationship is a two-way interaction that fully realizes the subject's potential (CIDEA 1997).

Thus the new model differs not only from conventional agronomy but from traditional teaching and learning methods as well. According to Elba Castro (1997), it is closely related to the educational principles of Paulo Freire, where the teacher must abandon language, descriptions, categories, and concepts in order to analyze the pupil's situation and relationships with the world. The use of this methodology trains the subject to understand a reality and to transform it, with the "educator" acting as facilitator. This educational process can only take place through participation in real-life activities, to then develop specific proposals or action areas with an integrated focus (Castro 1997). The goal of this methodology is not only to obtain new knowledge and skills but also to create new attitudes about nature and agriculture.

In Cuba, the practical development of sustainable agricultural training methods is both developed and implemented in the fields of small and large, state and private farms (most commonly cooperatives). This type of direct experience and experimentation has shown just how successful low-input sustainable agricultural methods can be, and has become the cornerstone in the teaching methodology for agroecology. In this context the Agroecological Lighthouses have proved extremely important. This practical aspect of development and implementation is fundamental to agroecological education. Thus, the principle of "learning by doing" is a vital aspect of the Cuban agroecological experience, and training programs generally require that approximately 50 percent of the total time schedule be spent in hands-on training activities.

Participants at the II International Course on Organic Agriculture analyze compost quality.

The use of participatory methods corresponds directly with agroecological principles. This not only means the use of active teaching methods with full student participation, but also an inherent holistic approach in this type of training. It would be nearly impossible to meet the challenge of educating millions of people by taking a conventional approach to agroecological education, especially given the innovative, constantly growing nature of this field. Accordingly it is felt that using participatory methods to generate and extend agroecological knowledge is the only solution to meet these educational needs in a reasonable time frame.

Cuban educators have used participatory methods in the past and experienced great success with them—starting with the "Popular Campaign Against Illiteracy" in 1961—in which thousands participated as volunteer teachers. Other such educational actions following the farmer-to-farmer methodology developed in Central America, combined with the basic educational principal of revolutionary Cuba that "those who know more teach those who know less," have been carried out. Projects of ANAP and the Council of Churches of Cuba in the city of Havana and Santa Clara province have trained hundreds of farmers, farm managers, and technicians using this approach.

Institutional Alliances for Agroecological Education

Agroecological education and training are intensifying in Cuba and will require increased collaboration among different organizations. While the leadership for university-level agroecological teaching rests clearly with CEAS, support from most of the universities and agricultural schools of the country has been crucial, along with support from the different branches of the Ministry of Agriculture, ANAP, and various research centers. The additional logistical, content development, and organizational support from the Organic Farming Group of ACTAF, and financial support from UN Development Program (UNDP), the Institute for Food and Development Policy (Food First), and Oxfam America have contributed to making this a highly successful program.

As in many fields in various countries, strategic alliances have been achieved among governmental and nongovernmental institutions to support agroecological education. Following the tradition of Cuban educators in terms of international solidarity, and of the international agroecological movement, Cuba has been cooperating with foreign organizations, farmer organizations and networks, NGOs, and academic institutions in different countries. Strong support for agroecological education in its initial stage in Cuba was received from the UN Food and Agriculture Organization (FAO) through a technical cooperation project (TCP) for training in sustainable agriculture in 1994–1995. Financial collaboration was also received from UNDP, the Latin American Consortium for Agroecology and Development (CLADES), Food First, and later from the German NGO, Bread for the World. Since at least 1993, Cuban professors, researchers, and producers in this field have contributed their experience and knowledge in Argentina, Chile, Uruguay, Bolivia, Brazil, Ecuador, Colombia, Peru, Venezuela, Mexico, Central America, and Spain and as well as in the US and Asia. Lectures, short courses, and four undergraduate and three master's level programs have been offered by Cubans at universities in these countries.

Hundreds of Latin American professionals attend practical training courses, and complete their bachelors, master's, and doctoral studies in agroecology in Cuba. More recently, university students from other countries take courses and do their graduate theses in Cuba. Thus hundreds of people learn from Cuban educators each year. Additionally, through focused delegations, hundreds of international agricultural professionals visit Cuba each year; this includes dozens of US farmers and professionals.

Conclusion

The agroecological training of technical staff has been an overwhelming task of strategic importance that has received a great deal of attention in Cuba. The high educational level of Cuban farmers and the extensive human resources and physical infrastructure for science has made rapid advances possible. Work carried out with the cooperation of universities, farmer organizations, government ministries, and NGOs has generated a rapid blossoming of agroecological consciousness in the agricultural sector, as well as a more gradual increase among the general public. The new pedagogical methods based on participation have been very satisfactory, and show promise of giving even better results in the future. An evaluation of these efforts leads us to conclude that while we have made tremendous progress, there is still a great need for this work to further embrace all sectors of society, furthering the conversion to sustainable agriculture in Cuba.

References

Altieri, M. 1995. *Agroecology: The Science of Sustainable Agriculture.* Boulder, CO: Westview Press.

Castro, E.A. 1997. Las dinámicas en la educación ambiental. *Resúmenes de la primera Convención Internacional sobre Medio Ambiente y Desarrollo.* Havana.

CIDEA. 1997. *Estrategia Nacional de Educación Ambiental.* Havana.

CLADES-FAO. 1991. *Informe de la reunión sobre curriculos in Agroecología y Desarrollo Rural Sostenible.* Santiago, Chile.

FAO-ORLAC. 1995. Desarrollo agropecuario. De la dependencia al protagonismo del agricultor. *Serie Desarrollo Rural* No. 9. Santiago, Chile.

FAO-ORLAC. 1996 *Rentabilidad en la Agricultura: ¿Con más subsidios o con más profesionalismo?* Santiago, Chile.

FAO. 1993. *Como mejorar la calidad de la capacitación.* Rome.

Funes, F. 1997. Grupo Gestor de Asociación Cubana de Agricultura Orgánica. *Resumenes del III Encuentro Nacional de Agricultura Orgánica.* Conferencias: 1–3. Villa Clara, Cuba.

García, L. 1997. La agricultura sostenible, la demanda de la educación superior y la formación de recursos humanos. *Taller de Desarrollo Sostenible en el Tropico Boliviano.* Universidad Autónoma Gabriel René Moreno. CIMAR-WWF: Santa Cruz, Bolivia.

García, L., N. Pérez, and P. Marrero. 1997. La conversión hacia una agricultura sostenible en Cuba. *Resúmenes del Congreso de educación ambiental para el desarrollo sostenible.* Havana.

Lane, P. 1997. El modelo cubano de desarrollo sostenible. *Seminario Internacional Medio Ambiente y Sociedad.* Havana.

Martínez, A. 1991. *La Planificación en la educación superior.* Universidad Estatal de Bolivar, Ecuador.

MINED. 1996. *Módulo de BME de los centros de Agronomía.* Havana.

Pérez, M. 1997. Programa de educación ambiental comunitaria en el medio urbano. *Resumenes del Congreso de educación ambiental para el desarrollo sostenible.* Havana.

República de Cuba. 1994. Plan Nacional de Alimentación. *Informe a la FAO y la OMS.* Havana, pp. 9–20.

Rosset, P.M. 1997. La crisis de la agricultura convencional, la sustitución de insumos y el enfoque agroecologico. *Agroecología y Desarrollo* No. 11/12:2–12. Chile.

Santa Cruz, G. and M. Mayarí. 1997. Aplicación de los principios de la agricultura y el desarrollo rural sostenible en los politécnicos agropecuarios. *Resumenes del III Encuentro Nacional de Agricultura Orgánica.* Conferencias: 56–58. Villa Clara, Cuba.

Sarandón, S. and G. Hang. 1995. El rol de la universidad en la incorporación de un enfoque agroecológico para el desarrollo rural sustentable. *Agroecologia y Desarrollo* No. 8/9:17–20. Chile.

UNESCO-OREALC and FAO-ORLAC. 1988. *Educación Básica y Desarrollo Rural.* Santiago, Chile.

Vandermeer, J., J. Carney, P. Gesper, I. Perfecto, and P.M. Rosset. 1993. Cuba and the dilemma of modern agriculture. *Agriculture and Human Values* 10:3:3–8.

PART II
Alternative Practices for a New Agriculture

CHAPTER SEVEN

Ecological Pest Management

Nilda Pérez, Center for the Study of Sustainable Agriculture, Agrarian University of Havana, and Luis L. Vázquez, National Plant Protection Institute

One of the key elements of the alternative agricultural model in Cuba is the reduction or elimination of synthetic pesticides in pest and weed management. Yet, contrary to popular opinion, the introduction of alternative pest management methods was not strictly due to the Cuban financial crisis of the 1990s. Integrated pest management practices have been established in Cuba since the beginning of the 1980s, primarily in response to the negative effects of intensive chemical use, and early biological pest control efforts go much further back.

The 1960s marked a new phase for plant protection in Cuba—as earlier the quantities of synthetic pesticides that were imported were negligible, and pest control strategies had been mainly based on cultural practices and the use of inorganic preparations. But extreme poverty of Cuban farmers at the triumph of the Revolution, and the fact that Cuba was primarily an agricultural country, made it evident that any proposal for development would start by solving the agrarian problem. On one hand, this required a change in the semi-feudal land tenure system, and on the other hand, it required progress toward a modern and efficient agricultural system—read "Green Revolution"—following dominant thought at that time (Montano et al. 1997).

As a result of the intensive agricultural systems that were developed then, new standards were established for the import and application of synthetic pesticides. In the 1960s and 1970s, pest control was based almost exclusively on the use of chemical pesticides. Predefined norms were used for each crop, which stipulated which chemicals to spray, at what dosages or rates, and when—based on calendarized applications.

During this phase of "chemicalization" of agriculture, cultural pest control practices and other traditional methods were largely abandoned by Cuban farmers, a trait demonstrated in many countries when synthetic pesticides first appeared on the world market. However, the consequences of heavy chemical use quickly became evident, as new pest and disease problems began to appear.

Concern grew with the appearance of new pests, the ineffectiveness of some pesticides in the control of existing pests, the development of pesticide-resistant insect populations, and decreased population densities of natural enemies of insect pests. This concern contributed to the creation of the State Plant Protection System in the mid-1970s by the Ministry of Agriculture (MINAG), and the subsequent development of Regional Plant Protection Stations (ETPPs) across the country. This new program was an early warning system based on nationwide pest population monitoring. The program was constantly fine-tuned, and was backed by a continuous training and education program. In 1975, the very first year of its existence, it reduced national pesticide consumption by half (see Figure 1).

Figure 1. Cuban pesticide imports since the initiation of pest population monitoring at Regional Plant Protection Stations

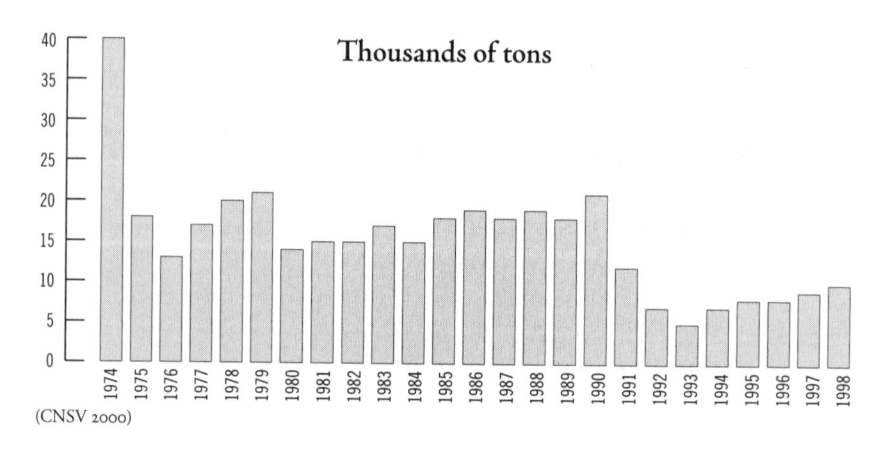

(CNSV 2000)

In order to build this network of ETPPs, a massive research program was mounted to study the biology, ecology, behavior, damage potential, thresholds, and control methods of the principle crop pests and diseases, which would give the early warning system a solid scientific base on which to stand. At the same time, this new body of technical knowledge constituted the foundation for future integrated pest management (IPM) programs. In 1982 IPM was established as the official policy of the Cuban state. Steps were taken to integrate cultural, chemical, and biological control methods in which the use of predators, parasitoids, pathogens, and antagonists were the most notable elements.

In the 1990s another remarkable reduction in pesticide use occurred, with a major difference with respect to 1974. By 1990 methodologies had been developed for mass production of biological control agents, the material basis was there, and so was the necessary organization. IPM programs had been in development since the 1980s, and were being used in 14 crops. As a result, between 1991 and 1998 pesticide imports fluctuated between just six thousand and twelve thousand tons, with an average of 8,375 tons per year (down from twenty thousand). This reduction has been more pronounced for insecticides, and to a lesser degree for fungicides and herbicides, because there were many more nonchemical control alternatives for insect control then for diseases or weeds. Today, in sugarcane, coffee, managed pastures, sweet potato, and cassava, synthetic insecticides are no longer used. In cabbage synthetic insecticide application is minimal or nonexistent, while use is also low in citrus, tobacco, plantain, and banana.

After the triumph of the Revolution, plant protection in Cuba passed through four stages, each one with a stronger agroecological emphasis than the one that came before (Vázquez and Almaguel 1997):

- diversification of land tenure and land use, and the diversification of agriculture (beginning in the 1960s)
- the creation of the State System for Plant Protection based on the ETPPs in the mid-1970s
- the implementation of a national program for biological pest control in the late 1980s
- the development of IPM programs with a focus on crop management in the 1990s

Recent developments in land tenure have reduced average farm sizes and increased the role of cooperatives in production, converting the countryside into a mosaic of crops, giving a boost to agroecological efforts.

In the establishment of a new agricultural model in Cuba, one of the most urgent tasks is to find ways to reduce synthetic pesticide use in pest management. Biological control is one of the best ways, and currently constitutes the main alternative available (Pérez et. al. 1995; Rosset and Benjamin 1994; Rovesti 1998). The use of natural enemies in Cuba dates back to the beginning of the century, but it was not until the 1960s that more complete programs were designed for their study and use, reaching massive and generalized levels in certain crops by the 1980s.

Biological Pest Control

The role of biological control in sustainable agriculture has been widely discussed. It has been amply demonstrated that the recovery of functional biodiversity in agroecosystems facilitates natural population regulation of pest species. In order to regain this natural regulation, the conversion process toward more organic farming must include management programs with strong ecological underpinnings, which promote the gradual recovery of lost biodiversity.

The Cuban model of conversion has been marked by an initial phase of input substitution, replacing chemical inputs with biological ones, within an IPM framework. In this phase there is an integrated combination of increased use of predatory insects and pest pathogens, with a more rational use of synthetic chemical pesticides. In these new programs, the use of resistant crop varieties, better agronomic practices, the management of pest habitats, and a concerted training effort among farmers, have opened up new opportunities for agroecological management.

Biological control is the cornerstone of the methodology behind implementing these management programs in Cuba. The release of pest predators, pathogens, and antagonists provides interim plant protection as a new ecological equilibrium becomes established. There are many examples in the literature on how, with increasing biodiversity, crop rotation, intercropping, and application of organic matter, the primary mechanisms of natural regulation that come into play are those of biological control.

The greatest successes have been obtained in the mass rearing and release of natural enemies and in the development, mass production, and application of insect pathogens. Cuba is presently among the leading countries in the world in production of biological agents for pest and disease control.

The first attempts in Cuba to manage pests by using natural enemies date back to 1930, when the yellow Indian wasp (*Eretmocerus serius*), a parasitoid, was introduced from Singapore for control of the citrus blackfly (*Aleurocanthus woglumi*), and to the use of *Rodolia cardinalis* for control of cottony-cushion scale (*Icerya purchasi*). Both of these insects were significant pests in citrus orchards, and the initial efforts were able to establish natural enemy populations (Vázquez and Castellanos 1997). This example of classical biological control is one of the most successful in the region, and since its implementation, there have been no serious outbreaks of either pest, except where related to the inappropriate use of synthetic pesticides.

Also in 1930, a program was set up for the rearing and release of *Lixophaga diatraeae*, an endemic tachinid fly which is a parasitoid of the sugarcane borer (*Diatraea saccharalis*). These initial biological control efforts in Cuba were later forgotten between the 1940s and the 1960s, as happened to biocontrol in most countries.

Nevertheless, the antecedents of the current program of biological control can be found in the old laboratories where, prior to 1959, *L. diatraeae* was mass reared for release in sugarcane. The six laboratories that were producing this fly in 1959 represented the beginning of modern biological pest control, and the model upon which small-scale regionalized production would later develop. Between 1960 and 1980, new technologies were developed and production levels were boosted in the six existing laboratories. In 1980, the Ministry of Sugar (MINAZ) created the National Program for Biological Pest Control. By the middle of the 1980s, biological control of the sugarcane borer had completely replaced the use of chemicals for this pest in Cuba. By 1995, MINAZ had built more than fifty small-scale regional Centers for Production of Entomophages and Entomopathogens (CREE), which produced 78 million flies annually that were released over 1.6 million hectares of sugarcane (Fuentes et al. 1998). In the 1990s these centers diversified to produce additional biological control agents, including entomopathogens (diseases of insect pests).

Thanks to advances in biological pest control and to the accumulated experience of the early CREEs (the majority created by MINAZ), by the mid-1980s a strong tendency developed to substitute these methods for synthetic insecticides. Contributing to these advances was the fact that as early as the 1960s the first commercial microbiological pest control products based on *Bacillus thuringiensis* (Bt) had begun to appear in the Cuban market. The introduction of some of these Bt formulas, combined with the success of the first trials for controlling the budworm (*Heliothis virescens*) on tobacco, and caterpillars (*Mocis latipes*) on pastures, stimulated interest in the search for native Bt strains. The results obtained in these early years, and knowledge gained from field experiments in the former USSR, demonstrated the feasibility of developing small-scale production technologies for entomopathogens.

A two-phase program was designed in the early 1980s based on these experiences. The first stage would develop small scale, semi-artisanal production technologies; the second stage would concentrate on the development of semi-industrial and industrial scale technologies. The second stage did not mean that semi-artisanal production was to be abandoned, but rather that both types of production would be necessary to meet Cuba's needs. In 1988, MINAG, following in the footsteps of MINAZ, approved the National Program for Biological Pest Control for three years, from 1988 to 1990, which was focused on the expansion of the network of CREEs (see Table 1). These in turn would offer products and services to state-run enterprises, cooperatives, and independent small farmers.

The creation of these laboratories is one of the most interesting aspects of pest management in Cuba. At the start of the 1990 financial crisis, this program allowed for the rapid substitution of imported chemical inputs with

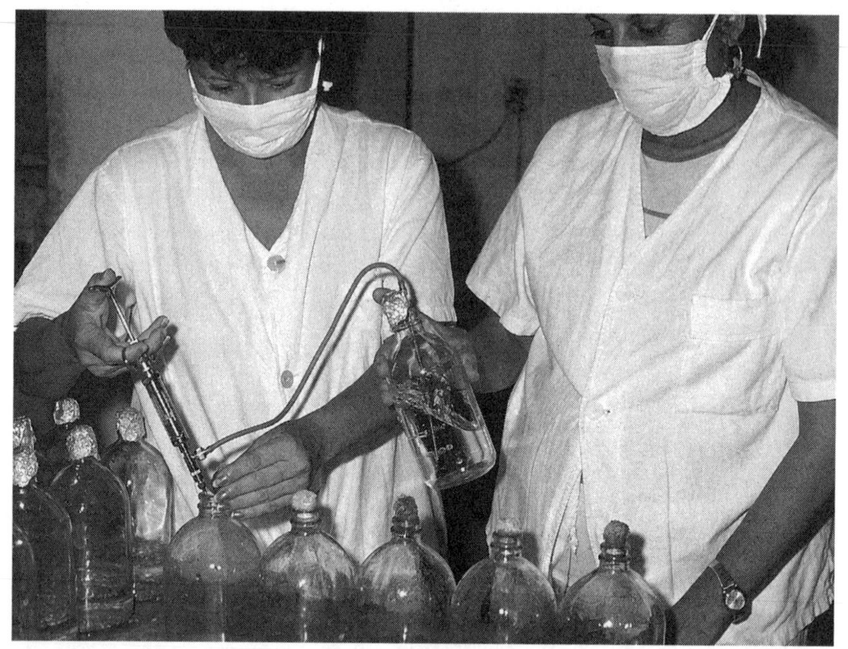

Inoculation of Beauveria bassiana *for production in liquid culture, at the CREE of "Amistad Cubano-Búlgara" Agricultural Production Cooperative.*

locally produced biological inputs, aided by the fact that small-scale production technologies were already well developed, and trained personnel were available. Additionally, farmers and producers had already recognized the advantages of these products; the combination of these factors resulted in the second major reduction in pesticide use (see Figure 1), and began a new era of plant protection in Cuba.

In 1991, the National Program for Biological Pest Control was reviewed at the highest levels of the Cuban state. As a result, MINAG and MINAZ agreed upon the creation of a total of 276 CREEs (222 in MINAG and 54 in MINAZ). In addition to these semi-artisanal centers, they planned to create 29 biopesticide plants with semi-industrial production technologies, as well as an industrial scale pilot plant for the production and development of new technologies. The numbers showed that the production capacity of these centers should cover the biological pest control needs of Cuban agriculture, allowing for the gradual reduction in pesticide use. The program has succeeded in creating 280 CREEs (four more than anticipated), three fermentation plants, and the pilot plant.

Table 1. Number of Centers for Production of Entomophages and Entomopathogens (CREEs) in non-sugarcane agriculture (MINAG), 1988 to 1999

YEAR	1988	1989	1990	1991	1992	1993	1999
Number	82	103	196	201	218	220	227

(CNSV 2000)

CREEs constitute the base of the national program for biological control in Cuba, and are considered a veritable revolution in the semi-industrial or artisanal production of biopesticides and natural enemies to control pests. The 280 CREEs are distributed throughout the country on state enterprises, cooperatives (CPAs), and the newly formed Basic Units of Cooperative Production (UBPCs). Fifty-three are located in sugarcane producing areas, and 227 in non-cane or fruit producing areas. They are positioned according to local needs and are established directly in agricultural areas or nearby. They have work teams comprised of university-educated specialists, lab technicians, and auxiliary staff. The decision as to the variety and quantity of organisms to be produced depends on the agricultural and livestock characteristics of each location. The products are sold directly to farmers and production units in the area, reducing transport and storage needs. Production is both highly diversified and specialized by region.

In addition to the diversity of organisms raised, a key aspect is that most of the pathogen strains selected for production are native, in accordance with the agroecological principal of utilizing local biodiversity. The search for and selection of new native strains and management techniques has been a permanent activity in this program. A good example of the importance of ecological principals in this effort is found in the use of three distinct varieties of Bt strains (see Table 2) in the control of the diamondback moth (*Plutella xylostella*), in order to retard the development of resistance.

Production methodologies have been established with flexibility, allowing for the use of the most appropriate and locally abundant materials as culture media within each region. For example, fruit juice is used as a liquid medium for Bt production. A variety of juices can be used, including orange, grapefruit, carrot, cucumber, sugarcane, etc. What is actually used in a given CREE depends on local availability—or more specifically, what can be salvaged from windfall, or other waste produce not meeting market quality. Similarly, rice, coffee, and sugarcane by-products are used for production of fungi. By utilizing by-products or wastes from agro-industrial processes, production costs are kept low. Business relationships are established directly between the farmers and their local CREE, which also offers technical advice, training, and extension services.

Studies on production costs by Sanchez et al. (1999) found that it is more expensive to produce entomopathogenic fungi than Bt bacteria, due to a longer incubation period and other factors. They also found that salaries represent from 27.5 to 46.5 percent of overall production costs in CREEs, and that quality control represents one of the largest expenses due to the relatively higher salaries and long hours dedicated to this activity. These authors determined production costs for 10 kg of the following products:

- Bt (*B. thuringiensis*), produced by liquid fermentation in static culture, 1.72 Cuban pesos[1]
- *Verticillium lecanii*, by liquid fermentation in static culture, 10.43 Cuban pesos
- *Beauveria bassiana*, produced by two-phase liquid-solid fermentation, 12.63 Cuban pesos

In terms of microbial biopesticides, CREEs now produce *Bacillus thuringiensis* bacteria, and *Beauveria bassiana*, *Metharhizium anisopliae*, *Verticillium lecanii*, *Paecilomyces lilacinus*, *Nomuraea rileyi*, and, recently, *Paecilomyces fumosoroseus* fungi, all for insect control. They also produce an antagonist fungus, *Trichoderma* spp., for the control of soil-borne pathogens. Following global trends, the most widely produced and applied biocontrol organism is Bt (more than 10,000 tons have been produced over 10 years).

The pests and diseases controlled with these products are documented in Tables 2, 3, and 4.

For Bt production, liquid fermentation in a static culture is used. This same technology was used for the production of various fungi, but due to the short shelf life of these formulations, it was replaced with solid or biphasic culture methods. Now, most fungi are produced through these technologies. In 1999, testing was carried out on the feasibility of producing Bt in solid culture, which clearly showed the advantages of liquid culture in this case.

One of the most important steps in the production process is quality control, for which a specialist is in each laboratory; there are quality control standards for each and every biological product. The final quality control assessment is made on two percent of daily production to determine purity, concentration, viability, and effectiveness in the field. There are also quality norms for the insects that are mass reared, such as *Trichogramma* spp., *L. diatraeae*, and *Sitotroga cerealella*. In addition, there is a government quality control system run by the network of Provincial Plant Protection Laboratories (LAPROSAVs) under the fiscal and regulatory control of the Plant Protection Institute (INISAV), that also provides certified microbial strains for the CREEs.

Table 2. Cuban strains of *Bacillus thuringiensis* and their use

STRAIN	PEST	DOSAGE*	CROP
Btk (LBT-24)	*Plutella xylostella* *Trichoplusia ni* *Erinnyis ello* *Spodoptera frugiperda* *Spodoptera* spp. *Ascia monuste eubotea* *Diaphania hyalinata*	4-5 l/ha	Vegetables Root crops Tubers
Btk (LBT-21)	*Heliothis virescens* *Plutella xylostella*	5–10 l/ha 1–5 l/ha	Tobacco Cabbage
Bt (LBT-13)	*Phyllocoptruta oleivora* *Polyphagotarsonemus latus* *Tetranychus tumidus*	20 l/ha 3–5 l/ha 5–10 l/ha	Citrus Potato, citrus Plantain
Btk (LBP-1)	*Plutella xylostella* *Mocis latipes*	5–10 l/ha 1–2 l/ha	Cabbage Pasture

*Refers to Bt produced in CREEs, as Bt obtained industrially has a higher concentration of crystals and requires smaller doses.

(Licor et al. 1995; Martínez et al. 1995; Carvajal 1995; Pérez 1996; Jiménez et al. 1997; and Fernández-Larrea 1999)

Mass production is not limited solely to the network of CREEs, but is also carried out at an industrial scale to produce significant amounts of *B. bassiana* in a concentrated powder formulation, and a liquid concentrate Bt product. As with the CREEs, the highest production volumes correspond to Bt. Four major strains are produced, three of them for Lepidoptera, and one for mites (Fernández-Larrea 1999).

Fungal-based biopesticides are also widely used, principally *B. bassiana* and *V. lecanii* (Table 3). There are several major uses of *B. bassiana*. It has been used very successfully in combination with pheromones in controlling the sweet potato weevil (*Cylas formicarius elegantulus*). *B. bassiana* is also applied to the soil to control the banana weevil (*Cosmopolites sordidus*), and for *Pachnaeus litus*. Recently it has also been used to control the leaf cutter ant (*Atta insularis*). *B bassiana* and *M. anisopliae* have also been used to control *Thrips palmi* in various crops. *Verticillium lecanii* has been applied with great success in whitefly (*Bemisia tabaci*) IPM programs for both tomatoes and beans, where it is used preventively.

Table 3. Use of entomopathogenic fungi to control insect pests

SPECIES	PESTS	DOSAGE	CROPS
Beauveria bassiana (LBB-1 strain)	*Cosmopolites sordidus*	1 kg/ha	Plantain
	Pachnaeus litus		Citrus
	Cylas formicarius		Sweet potato
	Lissorhoptrus brevirostris		Rice
	Diatraea saccharalis		Sugarcane
Verticillium lecanii (Y-57 strain)	*Bemisia tabaci*	1 kg/ha	Vegetables
	Myzus persicae		Fruit and root crops
Metharhizium anisopliae (LBM-11 strain)	*Mocis* spp.	5 kg/ha	Pasture
	Monecphora bicincta fraterna	5 kg/ha	Pasture
	Lissorhoptrus brevirostris	5–10 kg/ha	Rice
	Cosmopolites sordidus	20 kg/ha	Plantain
Paecilomyces lilacinus (LBP-1 strain)	*Meloidogyne* spp.	10–50 g/bag	Fruits, ornamentals, and root crops
	Globodera spp.	10–50 g	
	Rotylenchulus reniformis	50–100 g	
	Tylenchulus semipenetrans	10–50 g	
	Radopholus similus	50–100 g	
	Cactodera cacti	10–50 g/bag	

(López, 1995; Estrada and Romero 1995; Ayala et al. 1996; Pérez 1996; Trujillo 1997; Vázquez and Castellanos 1997; Rovesti 1998)

Farmers have widely adopted the use of *Trichoderma* antagonist fungi, usually applied directly to the soil to control soil-borne pathogens, but also applied to seedling roots at transplant, and to cuttings and seeds before planting (Table 4).

Table 4. The use of *Trichoderma* spp. to control soil-borne plant pathogens

ANTAGONIST	SOIL PATHOGEN	DOSAGE	CROP
T. harzianum (A-34 strain)	*Phytophthora capsici*	40 l/ha	Vegetables, ornamentals
	P. parasitica	40 l/ha	
	Rhizoctonia solani	40 l/ha	
	Pythium aphanidermatum	40 l/ha	
	Sclerotium rolfsii	40 l/ha	
Trichoderma spp.	*P. nicotianae*	40 l/ha	Tobacco

(Heredia et al. 1996; Pérez 1996; Rodríguez et al. 1997; Sacerio et al. 1997)

Figure 2. The production of entomopathogens and antagonists (metric tons) in CREEs and biological product plants

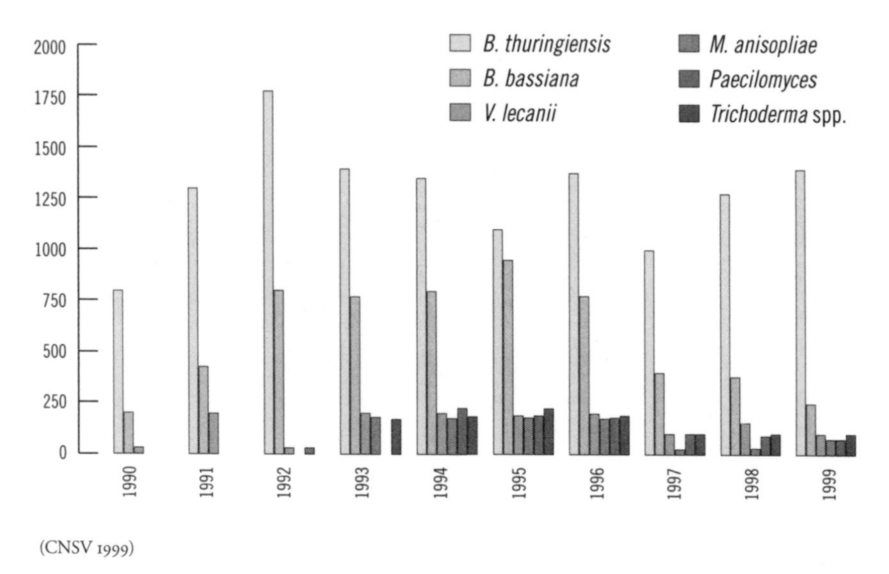

(CNSV 1999)

Figure 2 shows the combined production of entomophages and antagonists in CREEs and semiindustrial biopreparation plants from 1990 to 1999. Production levels respond to demand in each locality and year, and can fluctuate greatly. In 1990, 1,005 metric tons were produced, peaking in 1994 at 2,844 tons. Together with parasitoid rearing, coverage with biocontrol agents reached 982,000 hectares in the non-sugar sector in 1999.

It should be noted that with the increase in semiindustrial methods, improvements have been made in the formulation of the final product. Most importantly the production of liquid concentrates has increased, making it appear in Figure 2 as if Bt production has decreased in recent years. What has actually occurred is that an increasing percentage of the production is at a much higher concentration per liter, and thus less total product is produced, yet more hectares can be treated. In 1994 production began to be diversified by incorporating new microorganisms in the process, and more recently IPM programs have been improving and using other pest control methods. Because of this, pest management in many crops now relies upon multiple pest control methods, many of which are preventive rather than curative, as in the case of *Thrips palmi* and whitefly. In other words, pest management has evolved beyond the simple substitution of synthetic pesticides with biological ones.

Figure 3. Rearing of *Trichogramma* spp. (billions of insects/year) in MINAG CREEs

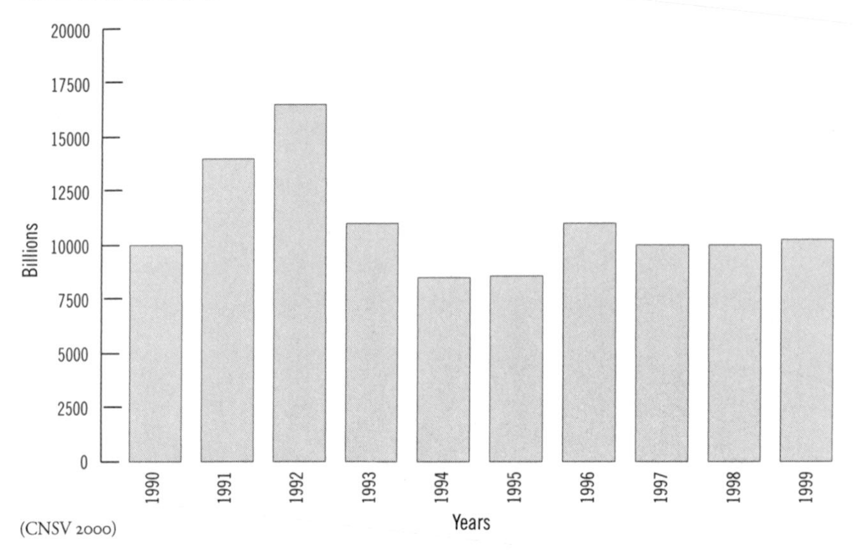

(CNSV 2000)

Figure 4. Production of parasitic nematodes (billions) for insect control in MINAG CREEs

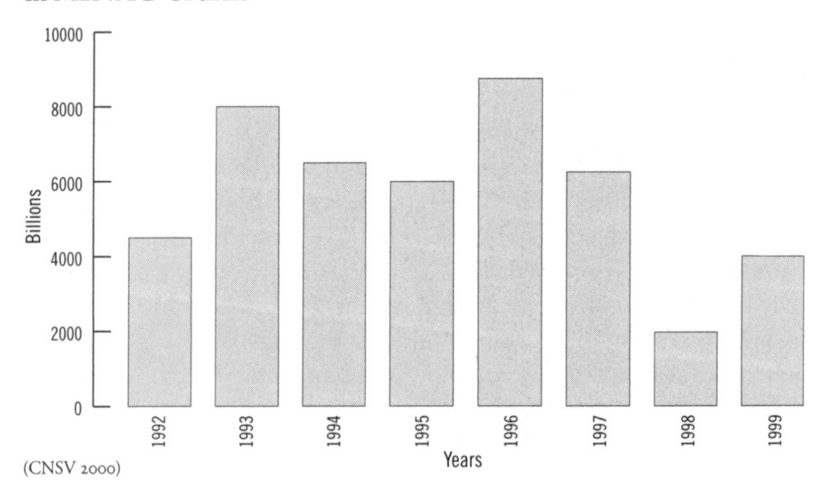

(CNSV 2000)

The beneficial insect that has been most widely reared and released is the egg parasitoid *Trichogramma* spp. (see Figure 3). Rearing uses *Corcyra cephalonica* or *Sitotroga cerealella* eggs as the host. The release is carried out when approximately 50 percent of the *Trichogramma* adults have emerged from their egg hosts in the CREE. Dosages range from 8,000 to 30,000 individuals per hectare, depending on the density of the pest to be controlled. The program is based on massive rearing of local biotypes. Table 5 shows the diversity of *Trichogramma* spp. and other beneficial insects being produced for the non-sugar sector. This egg parasitoid is of special interest due to its diversity and effectiveness in cassava and pastures. Key factors in success have been the management of local ecotypes and strict quality control systems.

Table 5. The use of beneficial insects in the non-sugar sector

CROP	PEST	BENEFICIAL INSECT	INUNDATIVE RELEASE	INOCULATIVE RELEASE
Pasture	*Mocis latipes*	*Trichogramma pretiosum*	X	
Cassava	*Erinnyis ello*	*Trichogramma pintoi* *Trichogramma* spp.	X	
Cassava	*Schizotetranychus caribbeaneae*	*Phytoseiulus macropilis*		X
Tobacco	*Heliothis virescens*	*Trichogramma* spp.	X	
Maize	*Spodoptera frugiperda*	*Telenomus* spp. *Euplectrus platypenae*		X X
Sweet potato	*Cylas formicarius*	*Pheidole megacephala*		X
Plantain	*Cosmopolites sordidus* *Tetranychus tumidus*	*Tetramonium guinense* *Phytoseiulus macropilis*		X X
Cabbage	*Plutella xylostella*	*Trichogramma* spp.	X	
Watercress	*Plutella xylostella*	*Trichogramma* spp.	X	
Cucumber	*Diaphania* spp.	*Trichogramma* spp.	X	
Beans	*Bemisia* spp	*Encarsia* spp.	X	

(Martínez and Pérez 1996; Pérez 1996; Vázquez and Castellanos 1997; Rovesti 1998)

The development of techniques for mass rearing and release of beneficial nematodes is still in the developmental phase. Satisfactory results have been achieved with *Heterohabditis* and *Steinernema*. Small amounts of these nematodes have been produced for citrus nurseries with the goal of controlling the citrus weevil *Pachnaeus litus* (see Figure 4).

Table 6 shows overall production levels in the CREEs and semiindustrial plants that produce biological control agents for sugarcane, which were used to treat 1.6 million hectares in 1995.

Traps with Galleria mellonella *parasitized by nematodes. Ministry of Agriculture, National Reference Center for Biological Production, Quivicán, Havana.*

Table 6. Biological control agents used in sugarcane

CONTROL AGENT	PEST	NO. OF CREES INVOLVED	PRODUCTION IN 1995
Lixophaga diatraeae	*Diatraea saccharalis* *Elasmopalpus lignosellus*	50	80.86 million
Trichogramma fuentesi	*Diatraea saccharalis* *Mocis latipes*	3	24.6 million
Eucelatoria sp.	*Leucania* spp. *Mocis latipes* *Spodoptera frugiperda*	6	2.0 million
Archytas monachi	*S. frugiperda*	3	0.28 million
Telenomus spp.	*S. frugiperda*	3	1.1 million
Euplectrus platypenae	*S. frugiperda*	3	0.26 million
Cotesia flavipes	*Diatraea saccharalis*	6	Experimental
Bacillus thuringiensis	Defoliators (Lepidoptera)	7	209 mil kg
Beauveria bassiana	*Diatraea saccharalis* Others	9	132 mil kg
Metharhizium anisopliae	Various root and stem pests	3	Occasional
Nomuraea rileyi	Lepidoptera	4	Occasional

(Fuentes et al. 1998)

Botanical Pest Controls

The use of plant extracts for pest control has been practiced since ancient times, and is part of agricultural traditions in many regions of the world. Many plants have naturally occurring secondary compounds that have a variety of functions, including serving as insect repellents and/or natural toxins. Many of these substances are broad spectrum and can affect both beneficial and harmful insects. Additionally, some botanical insecticides can be toxic to humans. Thus, there are associated risks which limit their use in organic production applications (Dudley 1988). However, through evolution plants have been engaged in the struggle against pests much longer than have humans, offering us a wide range of possibilities to choose from.

One of the most widely studied extracts in recent years is that obtained from the neem tree, *Azadirachta indica.* Its effectiveness has been demonstrated for insect, mite, and nematode control (Schmutterer and Ascher 1987; Hellpap 1989). One of the most beneficial aspects of the neem extract is that it only has a slight contact effect (the substance has to be ingested); thus, its effect on beneficial insects is limited, since they do not ingest the crop. Furthermore, the myriad of bioactive substances it contains minimizes the risk that resistance will develop, and it is not toxic to humans or other mammals. The chemical diversity of neem produces a range of actions against pest insects, including repellency, appetite suppression, sterilization, insecticidal action, and growth regulation.

To neem can be added many other exotic and endemic species represented in the Cuban flora, which produce substances with a regulatory effect on the populations of different harmful organisms. However, the use of botanical pest controls in national programs of plant protection was not considered until 1990 when research began in this field. To date, it has been observed that different extracts of approximately 40 species belonging to 25 families have biological activity on pest fungi, insects, mollusks, and rodents. Among the most important botanical families, due to the number of species studied which have given positive results, and also due to bioactivity and versatility, are the following: *Meliaceae, Asteraceae, Fabaceae,* and *Solanaceae.*

The most significant species are the neem tree, paradise tree *(Melia azedarach), Solanum mammosum,* and marigold *(Tagetes patula)* (Hernández et al. 1994). Current Cuban programs in development for extracting products from these plants include both research and the establishment of plantations to produce natural products for agricultural use. These plantations are designed to enhance the recovery of degraded marginal lands, generate biomass, and make ecological improvements (Estrada 1994). A project is in progress for the industrial development of the neem and paradise trees. This includes the establishment of 15 plantations of 12 hectares each (six hectares of neem and six of paradise), four processing plants (semiindustrial) with a capacity of 200 tons a year each, and a pilot plant for industrial production of commercial products. Presently, there are more than 300,000 trees, of which 25 percent have reached productive maturity. They should produce 2,500 tons of seeds in 1999 (Estrada et al. 1998).

One of the advantages of neem extract is that complex extraction techniques are not necessary, therefore, small-scale cottage industry production can be introduced to farmers and cooperatives. Small-scale production used for botanical control is one of the essential elements that make it sustainable. There are many ways to prepare neem extracts, but the recommended method for small-scale extraction starts with cleaning and drying the seed, followed by storage in a cool and dry place. When ready for use, the seed is ground into a powder that is applied in an aqueous extract. For this process, 20 to 25 grams of the powder are mixed with each liter of water and left to brew between six and eight hours before it is filtered.

Field application of neem extract is done in the afternoon, using a volume of final solution of 300–600 liters per hectare, requiring between 6 and 7.5 kg of ground seed. The powder can also be applied directly to the soil for nematode control at the rate of 100 grams per square meter (Estrada 1995). Research has shown that natural products based on neem are effective for the regulation of many insects, mites, and nematodes that affect vegetables. Thus, they are recommended for use in many crops. Table 7 shows pests that can be controlled by products with a neem base in 14 different crops, among them rice, tomato, corn, and beans.

The use of neem extracts is not limited to agricultural pest control. There are also effective neem based controls of ectoparasites in livestock, such as ticks (*Boophilus microplus*) and mites (*Meninia gynglimara*), and poultry lice (*Menophon gallinae*) on laying hens, as well as for the control of scabies on rabbits and pigs. Lice, fleas, and other ectoparasites of domestic animals have also been controlled with excellent results (Estrada and López 1998). As shown in Table 7 there is a group of natural products that will be increasingly available as plantations mature and enter full production. It must be made clear that these products are more preventive than curative. They are not a general solution and must be used only when they are absolutely necessary, and more importantly, within an ecologically managed program.

Table 7. Cuban neem products for pest control

CROP	PEST	PRODUCT	DOSAGE
Avocado	*Selenothrips rubrocinctus* *Psudacysta persicae*	OleoNeem 80 CE CubaNeem-SM	3 l/ha 7.5 l/ha
Peppers	*Bemisia tabaci* *Aphis gossypii* *Heliocoverpa zea*	OleoNeem 80 CE CubaNeem-SM	1.5 l/ha 6 kg/ha
Garlic	*Eryophis tulipae* *Thrips tabaci*	OleoNeem 80 CE CubaNeem-SM	1.5 l/ha 6 kg/ha
Rice	*Nezara viridula* *Spodoptera sunia* *Diatraea lineolata*	OleoNeem 80 CE OleoNeem 50 CE	1.5 l/ha 3 l/ha
Eggplant	*Thrips palmi*	CubaNeem-SM	6 kg/ha
Onion	*Thrips tabaci* *Eryophis tulipae*	OleoNeem 80 CE CubaNeem-SM	1.5–3 l/ha 6 kg/ha
Citrus	*Phyllocnistis citrella* *Toxoptera aurantii* *Aphis spiraecola* *Phyllocoptruta oleivora*	OleoNeem 80 CE CubaNeem-SM CubaNeem-T	1.5–3 l/ha 6 kg/ha 7.5 kg/ha
Cabbage	*Plutella xylostella*	OleoNeem 80 CE CubaNeem-SM CubaNeem-T	1.5 l/ha 6 g/ha 7.5 kg/ha
Beans	*Bemisia tabaci* *Empoasca kraemeri* *Diabrotica balteata*	OleoNeem 80 CE CubaNeem-SM CubaNeem-T	1.5–3 l/ha 6 kg/ha 7.5 kg/ha
Corn	*Spodoptera frugiperda* *Aphis maidis* *Helicoverpa zea*	OleoNeem 80 CE OleoNeem 50 CE CubaNeem-SM CubaNeem-T	1.5 l/ha 3 kg/ha 6 kg/ha 7.5 kg/ha
Melon	*Diaphania hyalinata*	OleoNeem 80 CE OleoNeem 50 CE CubaNeem-SM	1.5 l/ha 3 l/ha 4.5 l/ha
Cucumber	*Diaphania nitidalis*	OleoNeem 80 CE OleoNeem 50 CE CubaNeem-SM	1.5 l/ha 3 l/ha 4.5 l/ha
Tobacco	*Heliothis virenscens* *Bemisia tabaci*	OleoNeem 80 CE CubaNeem 80 CE	1.5 l/ha 6 kg/ha
Tomato	*Bemisia tabaci* *Keiferia lycopersicella* *Helicoverpa zea*	OleoNeem 50 CE OleoNeem 50 CE CubaNeem-SM CubaNeem-T	1.5 l/ha 3 l/ha 6 kg/ha 7.5 kg/ha

(Estrada and López 1998)

In 1989 and 1990 simple methods to extract nicotine from tobacco residues (a by-product of the cigar industry) were developed in an effort to control crop damage due to the whitefly. As a result, a new product commonly known as *Tabaquina* was developed and produced on CPA cooperatives and state enterprises that guarantees highly effective control of this and other pests. Tabaquina is now used throughout the country, and although it is considered a highly toxic and broad-spectrum insecticide, its half-life in the field is only four days.

Conservation and Management of Natural Enemies

Cuban agriculture is in transition from chemical pesticide use through the phase of input substitution. Early advances in biological pest control made it possible in the 1980s to move beyond the notion that chemicals are the centerpiece of pest control (Pérez et al. 1995). But it should be clear that biological agents do not provide the only way to substitute for chemical use. Only through IPM programs that include a broad array of ecological practices will it be possible to establish population regulation with an appropriate equilibrium between natural enemies, pests, and crops.

With respect to sustainable agriculture, research on biological controls must emphasize conservation strategies. Unfortunately, in recent years the world trend has been to investigate new biological control agents that can be formulated as commercial products—that is, stored, sold, and applied as though they were chemical pesticides. The "substitution of agrochemical inputs by other low-energy, biologically based alternatives," is one of the phases in the conversion process from conventional to sustainable agriculture (Altieri 1994). The truth is that in this process, biological products have a role, but additional pest management activities must be based on natural population regulation, in which native natural enemies play a significant role.

In classical biological pest control models, releases of beneficial insects are directed at the control of single pest species. However, the conservation and enhancement of existing natural enemies is a better preventive strategy, congruent with natural regulation of overall pest and plant pathogen populations in an agroecosystem. This is the strategy with the most potential in the context of sustainable agriculture. It is a matter of creating favorable conditions for the regulatory activity of local natural enemies, which incidentally also enhances the establishment and efficacy of introduced organisms.

An illustrative example of natural enemy conservation and management in Cuba is the case of the lion ant, *Pheidole megacephala*, a predatory ant whose action can be enhanced to control the sweet potato weevil. The successful enhancement of this native ant demonstrates the great potential that exists to propagate and use predatory ants in annual crops, as had been suggested by Carroll and Risch (1990) for perennial or long cycle crops. Cuba is the first country to develop practical procedures for the use and dissemination of ants for insect control in annual and semiperennial crops.

This management system is based on establishing natural reservoirs in areas where the ant population is abundant. *P. megacephala* colonies are transported from the reservoir to sweet potato fields, using traps that are made with banana stalks or leaves, burlap sacks, or dried coconut husks. The traps are sprinkled with a sweet molasses solution and then are kept damp to attract the ants. Once the ants have invaded the trap, it is transported to the field. As the trap dries up in the sun, the ants leave and begin foraging in the sweet potato fields (Castiñeiras 1986). The technique is widely used, with specific release rates for given regions. This same procedure is used for the control of the banana weevil (*C. sordidus*) with *P. megacephala* and *Tetramoniun guinense*, another predatory ant. In 1999, 8,470 hectares were treated with the ant management program, demonstrating its importance and potential.

As conditions are established which promote the activity and production of natural enemies, natural population regulation begins to come into play. This was demonstrated in a recent case study at the Jorge Dimitrov CPA, an Agroecological Lighthouse in the UNDP-SANE program. During the 1995–1996 season, there was an intense aphid (*Brevicoryne brassicae*) attack in a cabbage field (which had been preceded by maize in association with *Canavalia ensiformis*). The aphid population explosion came under natural control within a few days, by the parasitoid *Diaeretiella rapae* along with pathogenic fungi of the *Entomophtora* and *Erynia* genera (Pérez et al. 1998). Because chemical applications had been eliminated in this field, the agricultural ecology favored populations of these organisms, such that when the pest problem appeared all the farmers had to do was wait and watch. In years past, no farmer seeing such an infestation would have resisted the urge to spray chemical insecticides.

Crop Rotation and Intercropping for Pest Management

Cultural practices for pest control, such as crop rotation and intercropping, are key measures for ecological management. In Cuba, as in many other countries, cultural crop management methods for pest control were abandoned with the development of the high-input agricultural model. Although IPM programs have been promoted and implemented for more than decade in Cuba, only recently have other ecological management alternatives been considered to complement what had been achieved in the field of applied biological control.

Crop rotation is among the cultural practices that can be used to make the environment less favorable for the development of pests, disease, and weeds. Lampkin (1990) asserted that crop rotations are *the* most important measure for weed, pest, and disease control in organic production systems. Despite the fact that crops have been rotated since time immemorial, and the practice appears to be simple and hardly spectacular, it was one of the first practices dropped with the advent of the conventional model of industrial-style production. Thus, when the availability of synthetic pesticides and inorganic fertilizers decreased, it was one of the first practices to be rescued by *campesino* farmers. It should be noted that it was precisely in the small farm sector where the recovery in production began after the onset of economic crisis, responding to the abrupt fall in yields of many crops.

In some crops it was evident that rotation was a key element in the integrated management of certain pests, such as the case with nematodes in tobacco and the sweet potato weevil (Fernández et al. 1990, 1992). An IPM program in sweet potatoes was established for the weevil, which uses a rotation with at least two years between sweet potato plantings. Crop rotation is now one of the principle measures in the management of this pervasive and specialized insect that had been so difficult to control.

In weed management, even with the most persistent weeds, where other methods fail, rotation can be a very effective tool, as in the case of *Orobanche ramose.* In order to manage weeds through rotations, an important aspect is the competitive ability against weeds of each crop in the rotational sequence. Plant rotation with plants of high competitive ability can decrease weeds as effectively as seven or eight manual weedings. Producers know well that when sweet potato precedes another crop, weed incidence in the latter is much lower. Once the sweet potato canopy closes over the space between rows, weeds are completely shaded out. For example, *Sorghum halepense* is controlled through rotations of sweet potato with potatoes, beans, and peanuts.

Paredes (1999) recommends that crop rotations in areas that are highly infested with weeds should incorporate crops that are fast growing and that provide a high degree of shade coverage. He also argues that in order to take full advantage of the competitive and allopathic attributes of each crop in the rotation, the planting sequence is very important. For example in the control of *Sorghum halepense* and annual grasses the following planting schemes are useful: sweet potato–potato–legume–potato; sweet potato–potato–sweet potato–potato. To control purple nutsedge, *Cyperus rotundus,* the following rotations can be used: maize–potato–sweet potato–bean; maize–legume–sweet potato–bean; maize–bean–sweet potato–potato; sorghum–legume–sweet potato–bean. To control mugwort (*Parthenium hysterophorus*) and other broadleaf annuals the most effective rotation was found to be maize or sorghum–potato–maize–sorghum.

The effectiveness of a rotation depends upon many things, including which organism is to be regulated. The greatest successes have been attained against weeds and the nematodes that attack plant roots. Thus, crop rotation is used in Cuba in alternative weed management programs, for nematode control and to a lesser extent, for insect and disease management.

With extensive experience in the area of nematode control in Cuba, 53 varieties of 23 crops have been identified as resistant to species and strains of nematodes in the *Meloidogyne* genus. These resistant varieties can be included in rotation programs in infested areas, in open field conditions as well as in raised bed and urban production (Fernández et al. 1998).

Table 8 summarizes some of the research results obtained in recent years. In all cases, the main crop reaped benefits from a reduced nematode infestation index and/or lower weed populations.

Table 8. Effect of rotations on the management of soil pathogens, pests, and weeds

PRINCIPAL CROP	CROP IN ROTATION	PESTS CONTROLLED	REFERENCES
Tobacco	Peanuts	*Meloidogyne incognita* *M. arenaria*	Fernández et al. 1990
	Corn	*M. incognita* *M. arenaria*	
	Millet	*M. incognita* *Cyperus rotundus*	
	Beans–velvet bean	*M. incognita* *Eleusine indica* *Rottboellia exaltata*	
Potato	Cabbage–sweet potato	*M. incognita*	Gandarilla 1992
Potato	Sweet potato–beans–corn Beans–corn–sweet potato Corn or sorghum	*C. rotundus* *C. rotundus* Annual broadleaf weeds	Fernández et al. 1992
Tomato	Sesame	*Sorghum halepense* *M. incognita*	Fernández et al. 1992
Beans	Corn intercropped with velvet bean	*M. incognita*	Cea and Fabregat 1993
Corn	Peanut	*M. incognita*	Rodríguez et al. 1994
Soya	Potato–corn–potato Potato–sweet potato–potato	*Sclerotium rolfsii*	Hernández et al. 1997
Vegetables in organoponics	Onion or cowpea	*M. incognita*	Rodríguez 1998

The history of intercropping is similar to that of crop rotation. With the modernization of agriculture the monoculture system was intensified and extended. It is well known that the increase in pest problems is related to this expansion of monoculture, as the loss of biodiversity can be extreme in this model (Altieri and Letourneau 1982; Altieri 1995). Hence, one of main measures to be introduced into an agroecological management program is the elimination of monoculture as the basic structure of the agricultural ecosystem. For this to occur, diversification strategies must be defined, and intercropping is a key element within these strategies.

In Cuba's present agricultural reality intercropping presents a challenge. Increasing specialization of production units and increases in average field size have historically promoted large-scale monocultures. Crop associations (intercrops) were once prevalent, especially on small farms that required the maximum use of available land. This was true even in small-scale sugarcane production. Medium and large-scale producers used to allow field workers to intercrop in the sugarcane field as partial payment for activities such as weeding. In this manner, the worker was allowed to plant beans and/or peanuts in the spaces between the rows, and they were only paid with money for any weedings after they harvested. This traditional practice of intercropping food crops with sugarcane gradually disappeared, until food production became scarce in the beginning of the 1990s, when a small group of farmers recovered it, as in the case of crop rotation.

In recent years intercropping has come back into widespread use among small farmers or cooperatives, and in the ubiquitous urban organoponics (raised bed organic vegetable production). In an ethnoecological study of polycultures carried out in the agricultural communities of El Salvador (a mountainous municipality in the province of Guantánamo), crop associations were evaluated according to their productive and environmental performance and their acceptance by producers. Thirty-nine crop associations were used by the farmers of the region. Among the most frequent ones were coffee agro-forestry systems and associations of corn–beans and cassava–beans (Ros 1998). Pest populations are kept at such low levels by these intercrops that they do not constitute a problem in the region. Researchers from several scientific, educational, and production-oriented institutions are now involved in research to confirm this. Table 9 summarizes some of the results obtained.

In many cases, the pest control effect of crop associations works through natural enemy populations. Population increases of predatory insects and parasitoids are the most remarkable elements in common in these associations. The plant that is most extensively used as a secondary crop is maize. In the scientific literature, the role of this crop in attracting and supporting natural enemies has been well documented, as in the case of cassava (Pérez 1998).

Weed incidence can also be lower in intercrops. In general, an increase in the overall crop density decreases the ecological niche space open to weed infestations. The reduction of weed populations is greatest in associations with crops which show rapid growth and abundant foliage, thus shading out the weeds. This has been demonstrated in intercrops of cassava–maize, sugarcane–soybean, maize–bean and cassava–sweet potato (Leyva 1993). In maize–sunflower associations, the weeds *Brachiaria extensa, Digitaria decumbens* and *Echinochloa colona* are completely controlled (Pérez et al 1997).

In field trials conducted by Paredes et al. (1995), two different crop associations were designed with maize as the principal crop. One was designed for spring planting, made up of maize, squash, sweet potatoes, *carita* beans, and cucumbers. The other was designed for winter planting, and was composed of maize, tomato, beans, potatoes, and sweet peppers. In both systems favorable results were obtained for reduction of pests, diseases, and weeds.

Much more attention needs to be devoted to the role of intercropping in pest control. Little is known about the dynamics of pests, insects, diseases, weeds, and natural enemies in crop associations under field conditions. This is one of the main areas that should be studied in order to establish sustainable production systems. One of the best examples of the potential of intercropping for pest control is in urban agriculture, where small areas of diverse crops are planted. These urban gardens have reaffirmed that it is not necessary to use chemical pesticides in order to cultivate vegetables. Pest damage has been reduced using four basic strategies: design and preparation of the garden, agronomic management (Table 9), biological control, and training and public education (Vázquez 1995). Thus a national program has been implemented which has allowed for safe harvests without toxic residues.

Table 9. Intercrops that prevent pest outbreaks

SYSTEM	PESTS CONTROLLED	REFERENCES
Maize–velvet beans	*Meloidogyne* spp.	Cea and Fabregat 1993
Sweet potato–maize	*Cylas formicarius*	Suris et al. 1995
Cabbage–marigold Cabbage–sesame	*Bemisia tabaci* *Brevicoryne brassicae*	Vázquez 1995
Melon–maize Cucumber–maize	*Thrips palmi*	González et al.1997
Cabbage–tomato–sorghum–sesame	*Plutella xylostella*	Choubassi et al. 1997
Maize–squash–sesame Maize–squash–cowpea Maize–cassava–cucumber	*Spodoptera frugiperda*	Serrano and Monzote 1997
Squash–maize	*Diaphania hyalinata*	Castellanos et al. 1997
Tomato–sesame Cucumber–sesame	*B. tabaci*	Vázquez et al. 1997
Sweet potato–maize	*Cylas formicarius*	Quintero et al. 1997
Cassava–bean	*Erinnyis ello* *Lonchaea chalybea*	Mojena 1998
Cassava–maize	*E. ello* *L. chalybea*	Mojena 1998
Maize–tomato	*Bemisia* spp. *Liriomyza* spp.	Murguido 1995, 1996 León et al. 1998
Maize–bean	*B. tabaci* *E. kraemeri* *Aphis spiraecola* *S. frugiperda*	Murguido 1995, 1996 Pérez 1998
Cassava–maize–bean	*E. ello* *S. frugiperda*	Pérez 1998
Beans–sunflower	*Empoasca kraemeri* Chrysomelid defoliators	Alvarez and Hernandez 1997

Table 10. Cultural pest control practices used in urban agriculture in Cuba

PRINCIPAL PRACTICES	EFFECTS ON DIFFERENT PEST ORGANISMS
Selection of nematode-free site.	Avoids effects of *Meloidogyne* spp.
Leveling of the soil surface.	Reduces disease spread.
Use organic matter that is free of nematodes and weed seeds.	Avoids affects of *Meloidogyne* spp. and limits weed infestation.
Keep paths and surrounding areas free of weeds.	Limits pest access, reduces nematode and disease inoculant. Limits insect vector access and seedbed contamination.
Use pathogen-free seeds.	Reduces disease effects.
Plan adequate planting dates.	Limits pest incidence and avoids residual infestations.
Consider possible proximity effect if planting beds are too close together.	Avoids contamination of one bed by its neighbor.
Use a good planting density.	Reduces disease and pest losses.
Manual weeding.	Lessens effect of weeds.
Select and remove sick plants periodically.	Limits the spread of disease.
Remove and eliminate harvest residues.	Reduces sources of nematode inoculant, suppresses disease and weed levels.
Turn the soil in the planting beds twice in a 15-day period between June and August.	Reduces *Meloidogyne* spp. populations.
Use repellent plants intercropped in beds or planted along perimeters.	Forms a barrier that repels migrating pests.
Rotate crops (with garlic, onion, wild garlic, cabbage, sesame).	Reduces *Meloidogyne* spp. populations.
Intercrop with marigold.	Reduces *Meloidogyne* spp. and insect pests.
Solarization of beds before planting.	Lowers *Meloidogyne* spp. populations and weed densities.
Use tolerant varieties or those with partial resistance.	Reduces losses due to disease.
Good water management (avoid over-watering).	Limits disease incidence.

(Vázquez 1995)

Resistant Varieties

One of the primary strategies for pest management is the use of resistant, tolerant, or less susceptible varieties. The results of research on this topic have permitted the development of new varieties with these characteristics in various crops that are now being used, making a notable contribution to the reduction of pest and disease losses. A good example of this is the breeding and use of the Lignon tomato variety that is resistant to the geminivirus TyLCV strain (Gómez and Laterrot 1995). Table 11 demonstrates that this strategy is being intensely developed in Cuba as an alternative pest management method.

Table 11. Research results on resistant varieties

CROP TYPES	PEST OR DISEASE TOWARD WHICH RESISTANCE OR TOLERANCE IS EXHIBITED	REFERENCES
Vegetables Grains Oilseed crops Forage species Roots and tubers Tobacco Sugarcane Fruit trees Coffee	*Meloidogyne* spp.	Fernández et al. 1998
Melon	*Erysiphe cichoracearum*	Lemus and Hernández 1997
Tomato	*Stemphylium* *Fusarium* *Phytophthora* *Alternaria* *Meloidogyne*	De Armas et al. 1997
Tobacco	*Peronospora tabacina* (R) Tobacco mosaic virus *Phytophthora parasitica* *Rhizoctonia solani* (T)	Espino et al. 1998

Integrated Pest Management Programs

The current phase of pest control in Cuba is based on IPM. Two critical factors have contributed to the acceptance and implementation of these programs. The first has been the success of the early warning system implemented during the 1970s and the fact that it was based on solid scientific research. The second has been the introduction of new biological control agents within the framework of nascent IPM programs, which has made them more effective.

In Cuba, pest monitoring was never a daily activity until the 1970s, when the current network of Regional Plant Protection Stations (ETTPs) was developed and set in motion. At that time a new system of crop protection was established, based on regular field observation, evaluation of infestation levels, and advice to farmers on when to apply or not apply pesticides. This system is now used in the more than 27 crops which have established monitoring and advisory systems, for a total of 74 insect and mite pests, and several others for fungal diseases (Murguido 1987).

Today, a variety of IPM programs are used throughout the country in which a strong emphasis on the integration of nonchemical methods can be observed (see Table 12). In some of them, such as coffee, sugarcane, sweet potato, and cassava, no insecticides whatsoever are used for the control of insect pests.

Table 12. Level to which different management strategies are employed in IPM programs

CROP	CHEMICAL INSECTICIDES	BIOPESTICIDES	RELEASE OF NATURAL ENEMIES	CONSERVATION & ENHANCEMENT OF NATURAL ENEMIES	RESISTANT VARIETIES	AGRONOMIC PRACTICES
Coffee	N	L	N	H	M	H
Citrus	L	M	N	H	M	L
Sugarcane	N	L	H	H	H	M
Tobacco	L	M	N	L–M	M	M
Pastures	N	H	H	L–M	M	L
Rice	M	M	N	L–M	H	M
Corn	M	L	L	L	N	L
Beans	M	L	N	L	H	L
Cabbage	L–N	H	N–L	M–H	N	L
Tomato	M	M	N	L	H	M
Sweet potato	N	H	H	L	M	H
Cassava	N	M	H	H	M	H
Potato	M	H	N	M	H	M
Plantain and banana	L	M	M	M	H	H

Level of use: N = none; L = low; M = medium; H = high

Economic Benefits

By examining the economic data—without looking at noneconomic benefits—we find that the artisanal production and use of biological control agents saves Cuba hundreds of thousands of dollars annually (see Table 13). If we take into account the other critical uses that Cuba has for her scare foreign exchange—like importing medicines—we get a clear picture of what this means.

According to Maura (1994), the total cost of using biological controls on the crops studied came to 1,172,495 pesos. The purchase price for the toxic chemicals that would be needed to do the same job would be US $6,175,345. The job was done with biological controls at a tremendous savings in hard currency that would have gone to the international pesticide cartel. In contrast, most of the money spent on biological control stayed in Cuba, and it was almost all spent in pesos. Beyond that, the long-term savings in public health costs of not using those pesticides are simply incalculable, and the social impacts are of equal importance. Agroecological IPM programs have made urban agriculture possible, now providing much of the nation's fresh fruits and vegetables, in densely populated areas where pesticide applications would expose great numbers of people to dangerous chemicals.

The savings in agriculture are not just short term, either. Thanks to IPM programs, the populations of native natural enemies are increasing in various crops, and each year fewer applications are needed. This is the case in coffee, sugarcane, citrus, and cassava (where conservation of natural enemies is a key activity).

Table 13. Cost of biological control versus chemical pesticides

CROP	BIOLOGICAL PRODUCT	COST (PESOS)	INSECTICIDE REPLACED	COST (US$)
Horticulture	*B. thuringiensis*	501,430	Thiodan	1,622,253
Assorted crops	*B. thuringiensis*	243,303	Carbaryl	800,521
Grasses	*B. thuringiensis*	59,080	Carbaryl	397,613
Assorted crops	*V. lecanii*	54,048	Tamaron	431,788
Plantain	*B. bassiana*	134,106	Carbofuran	1,680,760
Sweet potato	*B. bassiana*	878,863	Tamaron	926,790
Rice	*M. ansipoliae*	80,290	Carbofuran	247,245
Plantain	*P. lilacinus*	79,236	Carbofuran	41,375

(Maura 1994)

Still to Be Done

Many of the elements already in place in Cuba's plant protection system are precisely those needed to ease the transition to a new form of agricultural production that is more sustainable. In spite of all that has been accomplished, much remains to be done to lessen chemical dependence.

Among our current priorities are:

- Investigate and implement the best possible diversification strategies, which are most in tune with local needs and realities, minimizing the planting of monocultures.
- Study insect pest, mites, pathogens, and weed populations in polycultures. Examine the correlation between the lowering of pest and pathogen levels and increased production in intercrops.
- Place emphasis on cultural practices, especially crop rotation, minimum tillage, and planting dates to minimize pest and disease incidence.

- Continue to perfect production techniques for biological control agents.
- Look for ways to enhance conservation of natural enemies. In the future biological control research should emphasize this strategy.
- Study interactions between crop, weed, and pest populations. Little is known about the role weeds play in the conservation and enhancement of natural enemies, and/or in the proliferation of pests.
- Evaluate the possible negative effects of inorganic fertilizers and synthetic pesticides in provoking pest outbreaks.
- Evaluate the effectiveness of ground cover, whether living, dead, or consisting of organic mulch, in controlling pathogens, insects, and weeds.
- Continue research on botanical products. Extend studies of the paradise and neem trees, and look into the potential of marigolds.
- Conduct a comprehensive study of trap crops and repellent plants that can be planted in association with crops, or in living barriers, to deter pest invasion.

Conclusion

In Cuba today, we have a pest management policy which takes into account ecological, economic, and social aspects of pest control. Our current plant protection system is supported by solid science, in which ecology has the central role. In the 1970s, when many nearby countries were still rapidly increasing their consumption of pesticides, Cuban consumption was cut by 50 percent with the birth of our plant protection system.

Everyone asks what will become of ecological pest management in Cuba, as we emerge from the economic crisis of the early 1990s. As more foreign exchange becomes available for the purchase of pesticides on the international market, it seems logical to some that Cuba will return to an intensive dependence on chemical inputs. Moreover, some think the current program of accelerated reduction of pesticide use is simply a short-term, stop-gap answer to maintain production until pesticide imports are affordable once again. But others—and they are more than a few—have a very different analysis, looking seriously at economic, social, health, and environmental factors, and conclude that the agroecological IPM model developed to date is simply a better model.

Our ecological policies have been further cemented by the recent passage of the Environment Law (República de Cuba 1997), in which Title 9, Article

132, is devoted to sustainable agriculture. With specific reference to pest management, it calls for:

> *b) The rational use of biological and chemical methods in accordance with local resources, characteristics, and conditions, to reduce environmental contamination to a minimum...*
>
> *d) Preventive and integrated management of pests and diseases, with special attention on the use of diverse biological resources.*

In light of recent history, it is hard to believe that Cuba would return to the calendar sprayings of the 1960s and early 1970s, or even the dependence of the 1980s. It is much easier to defend the assertion that other developing countries—faced with their own economic crises—should adopt a model not dissimilar to the Cuban pest management model, as a way to save foreign exchange and limit the impact of agrochemicals on public health.

Note

1. As the English language edition of this volume was being prepared, the parallel market exchange rate for the Cuban peso was roughly twenty to the US dollar.

References

Altieri M.A. 1994. Manejo integrado de plagas y agricultura sustentable en América Latina. *Taller sobre manejo integrado de plagas en América Latina.* Quito, Ecuador.

Altieri, M.A. 1995. Rotación de cultivos y labranza mínima. *Agroecología. Bases científicas para una agricultura sustentable.* Berkeley, CA: CLADES, pp. 173–183.

Altieri, M.A. and D.K. Letourneau. 1982. Vegetation management and biological control in agroecosystems. *Crop Protection* 1:405–430.

Alvarez, U. and C.A. Hernández. 1997. Influencia del intercalamiento del girasol (*Helianthus annus*) con frijol (*Phaseolus vulgaris*) y tabaco (*Nicotiana tabacum*) en el estado fitosanitario de estos cultivos. *Resúmenes del III Encuentro Nacional de Agricultura Orgánica.* Villa Clara, Cuba: Universidad Central de Las Villas, pg. 65.

Ayala, J. L., S. E. Castro, and S. Monzón. 1996. Evaluación de las posibilidades de control de *Mocis latipes* con diferentes entomopatógenos. IV Encuentro Nacional Científico-Técnico de Bioplaguecidas. Havana: INISAV, pg. 42.

Carroll, C.R. and S.J. Risch. 1990. An evaluation of ants as possible candidates for biological control in tropical annual agroecosystems. S. R. Gliessman (ed.), *Agroecology* Vol. 78, pp. 30–46, New York: Springer Verlag.

Carvajal, J. 1995. Uso de *Bacillus thuringiensis* cepa LBT 13 en el control de *P. oleivora* en el cultivo de los cítricos. III Encuentro Nacional Científico-Técnico de Bioplaguecidas. Havana: INISAV, pg. 18.

Castellanos, J., M.L. Sisne, A. Villalonga, W.E. Choubassi, and M.A. Iparraguirre. 1997. Estudio de *Diaphania hyalinata* (L.) y su comunidad parasítica en la asociación calabaza–maíz. *Resúmenes del III Encuentro Nacional de Agricultura Orgánica.* Villa Clara, Cuba: Universidad Central de Las Villas, pg. 55.

Castiñeiras, A. 1986. Aspectos morfológicos y ecológicos de *Pheidole megacephala* (dissertation). Havana: ISCAH.

Cea, M.E. and M. Fabregat. 1993. Dinámica y distribución de *Meloidogyne incognita* en un esquema de rotación de cultivos. *Informe Anual de Investigaciones.* Havana: ISCAH.

Choubassi, W.E., J. Castellanos, and M.A. Iparraguirre. 1997. Control de *Plutella xylostella* (L.) en Ciego de Ávila. *Resúmenes del III Encuentro Nacional de Agricultura Orgánica.* Villa Clara, Cuba: Universidad Central de Las Villas, pp. 54–55.

CNSV. 2000. *Estadisticas Centro Nacional de Sanidad Vegetal.* Havana: MINAG.

De Armas, G., T. Díaz, and J. C. Hernández. 1997. Mejoramiento para la resistencia ante enfermedades de importancia económica en el cultivo del tomate en Cuba. *Resúmenes Taller Nacional de Producción Agroecológica de Cultivos Alimenticios en Condiciones Tropicales.* Havana: Liliana Dimitrova Horticultural Research Institute, pg. 17.

Dudley, N. 1988. *Maximum Safety: Pest Control and Organic Farming.* Bristol Soil Association.

Espino, E., X. Rey, L. A. Pino, G. Quintana, N. Peñalver, and C. Bolaños. 1998. Habana Vuelta Arriba: nueva variedad de tabaco negro para cultivo en las provincias centrales y orientales del país. *Resúmenes, Forum Tecnológico sobre Manejo Integrado de Plagas,* pg. 8.

Estrada, J. 1994. El Nim y el Paraíso en Cuba, su cultivo y explotación como insecticida de origen botánico. *Resúmenes del Segundo Taller Nacional de Plaguicidas Biológicos de origen botánico.* BioPlag 1994, April. Havana: INIFAT, pg. 34.

Estrada, J. 1995. Progresos del cultivo del Nim y las investigaciones como insecticida natural. Resúmenes del Primer Taller Internacional y Tercero Nacional sobre Plaguicidas biológicos de origen botánico. BioPlag 1995, April. Havana: INIFAT, pg. 139.

Estrada, J. and M.T. López. 1998. *El Nim y sus insecticidas; una alternativa agroecológica.* Havana: INIFAT.

Estrada, J., R. Ronda, R. Montes de Oca, A. Rodríguez, J.M. Dueñas, A. González, B. Castillo, L. González, and E. Alvarez. 1998. Situación actual y perspectivas del Nim y sus bioinsecticidas en Cuba. *Resúmenes del Forum Tecnológico sobre Manejo Integrado de Plagas.* Matanzas, Cuba. September, pg. 30.

Estrada, M. and M. Romero. 1995. *Beauveria bassiana:* Una alternativa para el control biológico de *Diatraea saccharalis* en la caña de azúcar. III Encuentro Nacional Científico-Técnico de Bioplaguecidas. Havana: INISAV, pg. 17.

Fernández E., A. Pérez, E. Lorenzo, and E. Vinent. 1992. Efectividad del uso del ajonjolí como cultivo intercalado contra *Meloidogyne incognita. Protección Vegetal* 7:39–42.

Fernández, E., H. Gandarilla, and E. Vinent. 1990. *Manejo integrado de las plagas del tabaco en plantaciones.* Programa de Tabaco. Havana: ACC.

Fernández, E., M. Pérez, H. Gandarilla, R. Vázquez, M. Fernández, M. Paneque, O. Acosta, and M. Bastarrechea. 1998. Guía para disminuir infestaciones de *Meloidogyne* spp. mediante el empleo de cultivos no susceptibles. *Boletín Técnico.* INISAV 4 (3):1–18.

Fernandez-Larrea, O. 1999. A review of *Bacillus thuringiensis* (Bt) production and use in Cuba. *Biocontrol News and Information* 20 (1).

Fuentes, A., V. Llanes, F. Méndez, and R. González. 1998. El control biológico en la agricultura sostenible y su importancia en la protección de la caña de azúcar en Cuba. *Phytoma* 95:24–26.

Gandarilla, H. 1992. *Uso de la rotación papa–col–boniato en el manejo de nemátodos.* Havana: LAPROSAV.

Gómez, O. and H. Laterrot. 1995. Esperanza del mejoramiento genético para la resistencia a geminivirus en tomate. *Memorias, IV Taller Latinoamericano sobre Mosca Blanca y Geminivirus.* Zamarano, Honduras. Ceiba 36 (1):122.

González, N., R. Avilés, B. Cruz, E. Sotomayor, M. Ruíz, and N. Ramos. 1997. Comportamiento de *Thrips palmi* Karny en policultivos de campo. *Resúmenes del Taller Nacional de "Producción Agroecológica de Cultivos Alimenticios en Condiciones Tropicales."* Havana: Liliana Dimitrova Horticultural Research Institute, pg. 59.

Hellpap, C. 1989 Insect pest control with natural substances from the neem tree. Conferencia presentada en Seventh Conferencia Científica Internacional de IFOAM I. Burkina Faso. January.

Heredia, I., C. Alvarez, M. López, and S. Monteagudo. 1996. Biocontrol con *Trichoderma* spp. de hongos asociados a semillas. IV Encuentro Nacional Científico-Técnico de Bioplaguecidas. Havana: INISAV.

Hernández, C.A., E. Quintero, and L. Herrera. 1997. Posibilidad del uso de la rotación de cultivos para disminuir el inóculo del hongo *Sclerotium rolfsii* Sacc. en el suelo. *Resúmenes del III Encuentro Nacional de Agricultura Orgánica.* Villa Clara, Cuba: Universidad Central de Las Villas, pg. 56.

Hernández, M., J. Estrada , and V. Fuentes. 1994. Potencialidades de la flora cubana como fuente de bioplaguicidas. *Resúmenes II Taller Nacional de BioPlag 1994.* Plaguicidas de Origen Botánico. Havana: INIFAT.

Jiménez, J., R. Fernández, A. Calzado, and M. Vázquez. 1997. Efectividad de biopreparados nacionales de *Bacillus thuringiensis* para la lucha contra *Heliothis virescens* en tabaco. V Encuentro Nacional Científico-Técnico de Bioplaguecidas. Havana: INISAV, pg. 58.

Lampkin, N. 1990. Rotation design for organic systems. *Organic Farming*. UK: Farming Press Books, pp. 125–160.

Lemus, Y. and J. C. Hernández. 1997. Comportamiento de variedades e híbridos de melón frente al mildiu pulverulento de las cucurbitáceas (*Erysiphe cichoracearum*). *Resúmenes Taller Nacional de Producción Agroecológica de Cultivos Alimenticios en Condiciones Tropicales*. Havana: Liliana Dimitrova Horticultural Research Institute, pg. 17.

León A., E. Terry, M.A. Pino, and M.E. Domini. 1998. Efecto del policultivo maíz–tomate en el comportamiento de plagas insectiles en tomate. *Resúmenes del Forum Tecnológico sobre Manejo Integrado de Plagas*. Matanzas, Cuba, pg. 6.

Leyva, A. 1993. Las asociaciones y las rotaciones de cultivo. *Primer Curso de Agricultura Orgánica*. Havana: ISCAH.

Licor, L., L. Almaguel, and A. López. 1995. Control de a ácaros en toronja y naranja con *Bacillus thuringiensis* (LBT 13) en Ciego de Ávila. III Encuentro Nacional Científico-técnico de bioplaguecidas. Havana: INISAV, pp. 2–3.

López, M. 1995. PAECISAV. Un novedoso nematicida biológico para el control de nematodos fitoparásitos. III Encuentro Nacional Científico-Técnico de Bioplaguecidas. Havana: INISAV, pp. 4–5.

Martínez, Z. and T. Pérez. 1996. Uso de Phytoseiulus macropilis para el control de ácaros. IV Encuentro Nacional Científico-Técnico de Bioplaguecidas. Havana: INISAV, pg. 6.

Martínez, Z, T. Pérez, L. Almaguel, R. Brito, A. Hernández, E. Peñate, J. Lluvides, and R. Marín. 1995. Utilización de la cepa LBT 13 y de Phytoseiulus macropilis para el control de Tetranychus tumidus. III Encuentro Nacional Científico-Técnico de Bioplaguicidas. Havana: INISAV, pp. 7–8.

Maura, J.A. 1994. *Producción de Biopesticidas. El caso de Cuba*. Informe del Taller Regional sobre Tecnologías integradas de producción y protección de hortalizas. Cuernavaca, Mexico: FAO, pp. 69–74.

Mojena, M. 1998. Arreglos espaciales y cultivos asociados en yuca (*Manihot esculenta* Crantz). Modificaciones en algunas variables del agroecosistema y su influencia en los rendimientos totales. Havana: UNAH.

Montano, R., N. Pérez, and A.M. Vizcaino. 1997. Los Plaguicidas en Cuba: ¿y en el futuro qué? *Conferencias del III Encuentro Nacional de Agricultura Orgánica*. Villa Clara, Cuba: Universidad Central de Las Villas, pg. 23–27.

Murguido, C. 1987. Pronóstico de plagas de insectos y a ácaros. Seminario Científico Internacional de Sanidad Vegetal. Havana: pp. 5–6.

Murguido, C. 1995. Estrategias para el manejo de plagas del frijol común. Havana: Departamento de Protección de Plantas. Centro Nacional de Sanidad Vegetal.

Murguido, C. 1996. Programa para el manejo de la mosca blanca en tomate. Departamento de Protección de Plantas. Centro Nacional de Sanidad Vegetal.

Paredes, E. 1999. Manejo agroecológico de malezas y otras plagas de importancia económica en la agricultura tropical. Curso sobre bases agroecológicas para el MIP. Matanzas, Cuba.

Paredes, E., E. Pérez, A. I. Elizondo, and J. Almandoz. 1995. Efecto de los policultivos en la lucha contra plagas. IX Forum de Ciencia y Técnica. Havana: INISAV-MINAG.

Pérez, D., I.R. Gutiérrez, D. Ortiz, and R. Jiménez. 1997. Estudio de la asociación maíz-girasol. *Resúmenes del III Encuentro Nacional de Agricultura Orgánica.* Villa Clara, Cuba: Universidad Central de Las Villas, pg. 29.

Pérez, L.A. 1998. Regulación biótica de fitófagos en sistemas integrados de agricultura-ganadería (master's thesis). Havana: UNAH.

Pérez, N. 1996. Control Biológico: Bases de la experiencia cubana. *Agroecología y Agricultura Sostenible, Módulo 2: Diseño y Manejo de Sistemas Agrícolas Sostenibles.* Havana: CEAS-ISCAH, pp. 122–128.

Pérez, N. 1998. Control Biológico: Bases de la experiencia cubana. *Agroecología y Agricultura Sostenible, Módulo 2: Diseño y Manejo de Sistemas Agrícolas Sostenibles.* Havana: CEAS-ISCAH, pp. 122–128.

Pérez, N., L. Angarica, E. Treto, J. Gómez, A. Casanova, and M. García. 1998. Regulación de plagas en un sistema de producción con bases agroecológicas. *Resúmenes del Forum Tecnológico sobre Manejo Integrado de Plagas.* Matanzas, Cuba, pg. 7.

Pérez, N., E. Fernández, and L. Vázquez. 1995. Concepción del Control de Plagas y Enfermedades en la Agricultura Orgánica. *Conferencias y Mesas Redondas del II Encuentro Nacional de Agricultura Orgánica.* Havana: ICA, pp. 48–55.

Quintero, P.L., A. Hernández, N. Pérez, and A. Casanova. 1997. Influencia de la asociación boniato-maíz en la regulación biológica del tetuán (*Cylas formicarius* F.). *Resúmenes del III Encuentro Nacional de Agricultura Orgánica.* Villa Clara, Cuba: Universidad Central de Las Villas, pp. 56.

República de Cuba. 1997. Ley No. 81 del Medio Ambiente. *Gaceta Oficial de la República de Cuba* (special edition). Number 7, pp. 47–96.

Rodríguez, F., J. Almandoz, and J. Jiménez. 1997. Efectividad de *Trichoderma harzianum* en la reducción de la incidencia de *Pseudoperonospora cubensis* en el cultivo del pepino. V Encuentro Nacional Científico-Técnico de Bioplaguecidas. Havana: INISAV, pp. 58–59.

Rodríguez, L., L. Sánchez, and I. Rodríguez. 1994. Efecto de diferentes sistemas de rotación de cultivos en papa sobre el índice de infestación de *Meloidogyne incognita. Cultivos Tropicales* (15):AS–P. 32.

Rodríguez, R.C. 1998. Posibilidades de control de *Meloidogyne incognita* en organopónicos utilizando medidas de combate no químicas. *Resúmenes del Forum Tecnológico sobre Manejo Integrado de Plagas.* Matanzas, Cuba, pg. 18.

Ros, L.E. 1998. Estudio etnoecológico de policultivos en comunidades del municipio El Salvador, Provincia de Guantánamo (master's thesis). Havana: UNAH.

Rosset, P.M., and M. Benjamin (eds.). 1994. *The Greening of the Revolution: Cuba's Experiment with Organic Agriculture.* Australia: Ocean Press.

Rovesti, L. 1998. La lotta biologica a Cuba. *Informatore fitopatologico* 9:19–26.

Sacerio, C., M. D. Ariosa, and J. L. Armas. 1997. Efecto de *Trichoderma* sp. en el control de enfermedades fungosas de la papa. V Encuentro Nacional Científico-Técnico de Bioplaguecidas. Havana: INISAV, pp. 60–61.

Sánchez, M., A. Camarero, and T. Pérez. 1999. Costo de producción de los bioplaguicidas por tecnologías artesenales. *Protección Vegetal* 14 (1):33–41.

Schmutterer, R. and K.R.S. Ascher (eds). 1987. Natural pesticides from the neem tree and other tropical plants. *Resumenes de la III Conferencia de Nim.* Nairobi.

Serrano, D. and M. Monzote. 1997. Policultivos con maíz en sistemas integrados ganadería-agricultura. *Resúmenes del Taller Nacional de "Producción Agroecológica de Cultivos Alimenticios en Condiciones Tropicales."* Havana: Liliana Dimitrova Horticultural Research Institute, pg. 79.

Suris, M. et al. 1995. Evaluación entomológica de once sistemas de rotación de ciclo corto para la papa. *Resúmenes del II Encuentro Nacional de Agricultura Orgánica.* Havana: ICA, pg. 65.

Trujillo, Z., 1997. Combate de la bibijagua (*Atta insularis*) con el insecticida biológica BIBISAV. V Encuentro Nacional Científico-Técnico de Bioplaguecidas. Havana: INISAV, pg. 64.

Vázquez, L. 1995. Efecto de siembras mixtas sobre plagas en "organopónicos." Informe interno. Havana: INISAV.

Vázquez, L. and L. Almaguel. 1997. Tendencia Agroecológica de la Protección de Plantas en Cuba. I Convención Internacional sobre Medio Ambiente y Desarrollo. Havana.

Vázquez L. and L. Castellanos. 1997. Desarrollo del control biológico de plagas en la agricultura cubana. *AgroEnfoque* 91:14–15.

Vázquez, L., D. López, and R. Rodríguez. 1997. Lucha contra las moscas blancas en los huertos urbanos. *Resúmenes del Taller Nacional de "Producción Agroecológica de Cultivos Alimenticios en Condiciones Tropicales."* Havana: Liliana Dimitrova Horticultural Research Institute, pg. 71.

CHAPTER EIGHT

Intercropping in Cuba

Antonio Casanova and Adrián Hernández, Liliana Dimitrova Horticultural Research Institute, and Pedro L. Quintero, Civilian Aeronautics Institute of Cuba

The economic crisis in Cuba has led to the implementation of agricultural alternatives long forgotten by conventional modern agriculture. Among these traditions and practices is the resurgence of intercropping—planting multiple crops in the same field in the same year. Under tropical conditions intercropping is a critical element in sustainable agriculture.

Before the triumph of the Revolution, more than half of the total area under cultivation was planted in vast monocrops of sugarcane, which far surpassed other crops such as tobacco, coffee, and fruit trees. Beginning in the 1960s, there were major changes in the Cuban countryside. Among these were the increasing intensification and specialization of production units. These changes brought higher yields and labor productivity to the principal nonsugar crops, especially on the large-scale state farms and the larger cooperatives. These farms were heavily mechanized, and relied on large-scale irrigation systems and a dramatic increase in the use of chemical fertilizers, synthetic pesticides, and herbicides. As more land was devoted to this type of production, Cuban farmers became more specialized, further ingraining monocropping as the dominant agricultural model.

The result of the increased use of chemical fertilizers, herbicides, and pesticides was a disruption of biodiversity, and the loss over time of soil fertility and productive capacity. By the end of the 1980s, outbreaks of diamondback moth *(Plutella xylostella)* severely damaged cabbage crops and other leafy greens, while other new pests like the whitefly (*Bemisia* spp.), affected beans, tomatoes, and other vegetables, with significant yield loss. At the root of many such problems was the extensive monocrop production system with its subsequent loss of biodiversity, much larger plot sizes, and high degree of specialization.

Intercrops or polycultures—which can be used to reduce monoculture and its negative impacts—can be defined as the production of two or more crops in the same area, during the same year. By sowing consecutively or in association agricultural production can be made more efficient by making better use of space and time (Leihner 1983). These are agricultural ecosystems that have varying degrees of complexity depending on which species are selected and how they combine with one another. When combined appropriately intercrops can be far more productive than monocrops and provide multiple benefits to the agroecosystem (Amador and Gliessman 1989).

From an ecological perspective, intercrops can reduce crop damage caused by pests through several mechanisms. One crop can function as a physical barrier to pest movement between the rows of another crop, and/or insect pests can be confused by the odors and colors of the different crops. Thus, the rate of pest invasion, dissemination, and reproduction can be lower in polyculture systems than in monocultures. One crop can also provide a habitat or refuge, and/or supplemental food source such as nectar, attracting beneficial insects that may be natural enemies of the pests of another crop in the same system. Crop rotations can help break the reproductive cycles of pests and disease by eliminating the host crop during one or more crop cycles. Furthermore, soil conditions can be improved when nitrogen-fixing legumes are part of the system, which, when they act as cover crops, can also reduce soil erosion and soil temperatures, helping to maintain critical soil biota. In many cases the added crop residues in an intercrop or rotation increase organic matter in the soil as well.

Intercropping has long been part of the agricultural landscape in many developing countries. In Latin America, more than 40 percent of cassava (*Manihot esculenta*), 60 percent of maize (*Zea mays*), and 80 percent of beans (*Phaseolus vulgaris*) are cultivated in intercrop systems (Leihner 1983).

The Origin of Intercrops in Cuba

Polycultures have been present in Cuba since pre-Columbian times. Their use grew in the early nineteenth century near the sugar mills where workers and slaves had self-provisioning areas, and in the *conuco* (farms of runaway slaves). During this period sugarcane areas had more agrobiodiversity; the big plantations had to feed their slaves and workers by planting food crops, and there were still many *campesinos* with diversified farms who also sold to the sugar mills. After the so-called Spanish-American War in 1898, sugar regions became much less diverse, as US sugar companies bought up most sugar mills, as well as the farmland previously owned by national producers and small farmers.

Cassava-corn polyculture, "Gilberto León" Agricultural Production Cooperative, Havana.

The new owners were interested only in producing sugar. Sugarcane took over the food-producing plots on the plantations, and day laborers were forced to look elsewhere for subsistence. However intercropping remained important in what was left of the subsistence and *campesino* sector—especially as it allowed for greater productivity from their small plots (Leyva 1993).

Traditionally small-scale producers have intercropped their sugarcane with short-cycle crops such as common beans, tomatoes, peanuts, and soybeans. Similarly, the open space between rows in plantain or fruit orchards was frequently used to cultivate annual crops. These intercrop systems had higher total productivity, greater crop diversity, and made more efficient use of available resources, with the added benefit of reducing pest, plant diseases, and weed levels, making them more profitable.

Historically, intercrop systems were not just developed by *campesinos* involved in a subsistence economy, but also by farmers in key productive regions like Baracoa and Maisí. The mountainous regions of eastern Cuba produce coffee, cocoa, and coconut (Hernández 1998). Here structurally complex agroforestry systems were developed, as in the case of traditional shade-grown coffee systems, which contribute to overall habitat improvement and increased biodiversity. These highland agricultural zones never passed through the phase of intensive agricultural farming techniques like the monocultures of "full-sun" coffee varieties found in other coffee-growing countries—thus these complex systems never disappeared.

Intercropping Today

The shortage of inputs for agriculture during the economic crisis has favored a return to intercropping techniques across Cuba. For example, today one finds combinations of vegetables with tropical roots, tubers, grains, or fruits. These crop combinations are based on the traditional knowledge of small farmers from different regions of the country. Today Cuban scientists are collecting and evaluating this empirical knowledge to adapt it to a variety of environmental conditions (Hernández 1998).

Intercropping and strip-cropping are the most widely used methods of combining crops. At present the most common crops being combined are cassava, maize, bean, tomato, sweet potato, squash, plantain, lettuce, radish, onions, sweet peppers, and other vegetable crops (see Tables 1 and 2).

Table 1. Common intercrops in Cuba

INTERCROPS	SOWING OR PLANTING TIME
Cassava–maize	simultaneous
Cassava–bush bean	+ 10 days for beans
Cassava–tomato	simultaneous
Cassava–cowpea	simultaneous
Cassava–tomato–maize	simultaneous cassava and tomato followed by maize
Maize–peanut	simultaneous
Maize–common bean–squash	+ 20 days for squash
Maize–*canavalia*	+20 to 30 days
Squash–maize	+20 days for maize
Sweet potato–maize	simultaneous
Sweet potato–sunflower	simultaneous
Plantain–beans	during plantain development
Plantain–peanut	during plantain development
Plantain–many vegetable crops	during plantain development
Taro–maize	simultaneous
Coffee–shade trees*	simultaneous
Cocoa–shade trees	simultaneous
Coffee–plantains–coconuts	simultaneous

**Piñon florido, Gliricidia sepium,* and other species.

(Quintero 1995; Leyva 1993; Hernández 1998; Mojena 1998)

Table 2. Common intercrops in urban organoponics (raised beds)

Intercrop
Lettuce–radish
Lettuce–chard
Lettuce–garlic or onions
Cabbage–lettuce
Cabbage–chard
Cabbage–garlic or onions
Pepper–radish
Pepper–garlic or onions
Pepper–lettuce
Pepper–chard
Pepper–string beans
Bush beans–lettuce
Bush beans–chard
Bush beans–garlic or onions

(Casanova and Savón 1995; Casanova 1995; Caraza et al. 1996)

Research Efforts

In previous decades researchers in Cuba did not study intercropping. More recently, however, because of the importance of this kind of low input sustainable agriculture to Cuban production, many new research programs, masters' theses and doctoral dissertations have focussed on the theory and practice of intercropping. There are also a number of participatory research programs involving farmers and researchers, designed to build a better understanding of diverse agricultural systems, and to find the best ways to promote them in the farming community. Efforts have been made to encourage farmers to rediscover traditional methods of intercropping, and to incorporate scientific advances from basic research. Thus many hybrid methods have been adapted to individual, cooperative, and state production systems, leading to more rational utilization of available resources. The main emphases of research have been on the development of crop associations in both time and space, the selection of genotypes that do well in these systems, the relationship between crop genotypes and the agroecosystem, soil conservation, pest regulation, and yield enhancement in intercrops.

Polyculture systems offer the farmer crop diversification, greater total productivity, and greater efficiency in land use. In order to evaluate the efficiency of an intercrop, we employ the land equivalent ratio (LER), which represents the relative area of land that would be needed in monocultures to give the same total production as when the crops are planted in association (Vandermeer 1989). This ratio is calculated with the following formula:

$$LER = LER_1 + LER_2 + \ldots + LER_n$$

where:

LER = is the cumulative LER of the system, and

$LER_1, LER_2, \ldots, LER_n$ = the individual LERs of each crop in the association which are obtained from the equation:

$LER = A_x/M_x$, where:

A_x = yield of crop x in the association

M_x = yield of x in monoculture

If the overall LER >1, then intercropping is more efficient than monocropping, if LER = 1, intercropping and monocropping produce the same results, while if LER < 1 monocropping is more efficient than intercropping.

In other words, the LER tells us how much total land would be required to produce the same quantity of each crop when planted separately, compared to how much is needed if they are intercropped. For example, empirical studies have shown that one hectare of cassava and tomato planted together produces the same total yield as 1.86 hectares of cassava and tomato planted separately (LER = 1.86).

In Cuba, most research projects on intercropping evaluate the LER (see Table 3) as one key indicator. Based on the LER rating, all of the common intercrop associations studied have proven to be more efficient than monocultures, by consistently reaching a LER of greater than 1. In combinations where cassava is used as the main crop, the LERs found by different researchers range from 1.60 to 1.98. Thus, the advantage of intercropping this long-cycle crop with short-cycle crops is clear.

Table 3. Sample land equivalent ratios (LER) obtained in Cuba

CROP ASSOCIATION	AUTHOR	LER
Cassava–common bean	(Mojena et al. 1996)	1.60
Cassava–common bean	(Hernández et al. 1997)	1.75
Cassava–maize	(Quintero 1998)	1.75
Cassava–maize	(Mojena 1998)	1.68
Cassava–maize	(Hernández et al. 1997)	1.82
Cassava–tomato	(Hernández et al. 1997)	1.86
Cassava–tomato–maize	(Quintero 1998)	1.98
Sweet potato–maize	(Quintero et al. 1997)	1.68
Sweet potato–squash	(Quintero 1998)	1.10
Lettuce–maize	(Quintero 1998)	1.67
Cucumber–lettuce*	(Caraza et al. 1996)	1.44
Cucumber–radish*	(Caraza et al. 1996)	1.93
String bean–radish*	(Caraza et al. 1996)	1.86

*In intensive raised bed systems, or organoponics.

Another crop with great potential for this type of system in Cuba is sugarcane. Its agricultural characteristics allow for symbiotic relationships with short-cycle crops such as legumes (if they are managed properly). Results from earlier test plots intercropping black beans and soybeans between sugarcane rows have yielded positive economic results. Recent research has demonstrated that under the climatic and soil conditions of the eastern part of Cuba, cowpea and peanut show satisfactory LERs when intercropped in sugarcane. However sunflower produced an LER of less than one both in new plantations as well as when planted in re-growth (Leyva 1993).

Some authors consider that other indices beyond the LER should be used, such as production in units of energy (MJ/ha), total protein yield (kg/ha) and profitability. The results reported in Tables 4, 5, and 6 indicate the potential benefits of intercrop systems as measured by these indices (Quintero 1999).

Table 4. Comparison of total energy production in intercrops and monocultures (MJ/ha)

CROPS	MJ/HA					DIFFERENCE BETWEEN INTERCROPPING AND MONOCULTURE
	INTERCROPS			MONOCROPS		
1 + 2	1	2	TOTAL	1	2	
Sweet potato + maize	72,000	27,800	99,800	64,800	41,700	35,000
Sweet potato + squash	69,600	360	69,960	78,240	1,740	(8280)
Beans + maize	8,580	12,232	20,812	8,580	31,970	12,232
Cassava + maize	88,550	31,970	120,520	93,500	38,920	27,020
Cassava + tomato	64,350	21,200	85,550	93,500	20,300	(7,950)
Lettuce + maize	8,680	29,190	37,870	10,150	36,140	27,720

(Quintero 1999)

Table 5. Comparison of total protein production in intercrops and monocultures (kg/ha)

CROPS	PROTEIN PRODUCED					DIFFERENCE BETWEEN INTERCROPPING AND MONOCULTURE
	INTERCROPS			MONOCROPS		
1 + 2	1	2	TOTAL	1	2	
Sweet potato + maize	270	180	450	243	315	207
Sweet potato + squash	261	4	265	293	20	(28)
Beans + maize	132	79	211	132	207	79
Cassava + maize	161	207	368	170	252	198
Cassava + tomato	117	212	329	170	203	159
Lettuce + maize	149	189	338	174	234	164

(Quintero 1999)

Table 6. Economic rate of return in intercrops and monocultures (%)

CROP	PROFITABILITY (CROP VALUE/PRODUCTION COSTS) X 100	DIFFERENCE BETWEEN INTERCROPPING AND MONOCULTURE
Sweet potato + squash	296	(7)
Sweet potato monoculture	303	–
Beans + maize	409	44
Beans monoculture	365	–
Lettuce + maize	220	101
Lettuce monoculture	119	–

(Quintero 1999)

Using Polycultures in Pest Management

Research and farmer experience have proven that the benefits of using various intercrop combinations include a reduction in weed infestation and pest attack. The advantages of intercrops in weed control have been demonstrated in Cuba for a variety of crops, particularly cassava (Hernández 1998; Mojena 1998). Hernández tested three cassava genotypes with three bean varieties with different growth habits, and observed a reduction of up to 70 percent of weeds when cassava was planted with bean varieties of indeterminate growth (Figure 1).

Figure 1. Reduction of weeds in bean/cassava polycultures

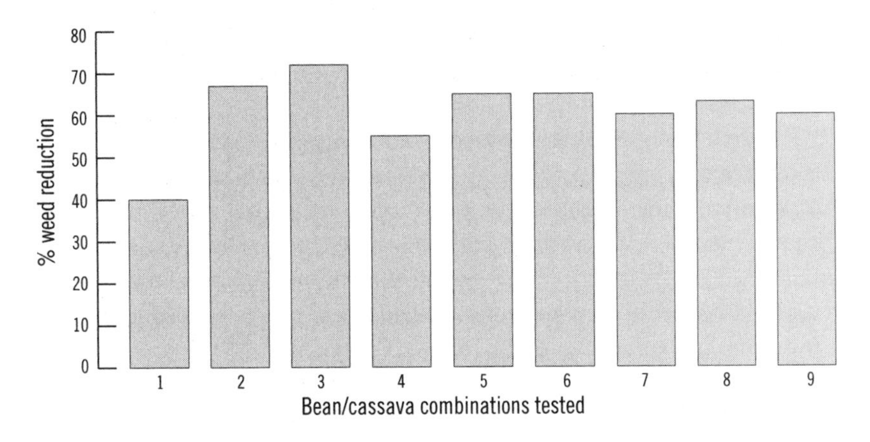

The lion ant, *Pheidole megacephala,* is considered an effective biological control agent for the sweet potato weevil. Vandermeer (1989) suggests that this predator is the operative mechanism of facilitation in the sweet potato–maize polyculture. Sweet potato, the principle crop, is facilitated by the positive effect of maize on the predatory ant. This polyculture has become very common in the municipality of San Antonio de los Baños in Havana province, thanks to the Agroecological Lighthouses set up by the Sustainable Agriculture Networking and Extension (SANE) project, and is now spreading to other farming regions of Cuba.

Lower indices of *Spodoptera* damage on maize have been observed in maize–bean polycultures than in monocultures. Because the beans rapidly cover the ground between maize rows, this insect finds it difficult to pupate, which it normally does on bare soil. Another successful crop combination is carrot–cabbage, where the repellent effect of the carrot substantially reduces diamondback moth damage on the cabbage (Santos, personal communication, 1995). The combination of cabbage, sorghum, and sesame was evaluated for diamondback moth control in Villa Clara province. The sorghum and sesame were planted as living barriers between every few rows of cabbage. Researchers have found an increase in natural enemy populations associated with a drop in diamondback moth and aphid populations in this system (Gómez Sousa 1999, personal communication).

With the recent generalized increase in knowledge of agroecology in Cuba, many ecological principles are now being widely applied, with a corresponding increase in farm- and plot-level biodiversity. Polycultures have proven to be an important tool in the context of IPM for pest-sensitive crops such as tomato and potato. This is the case with the IPM strategy to control the whitefly-gemini virus complex in tomato by using living barriers of maize in the exterior and interior of the seed beds and open fields. The maize barriers are planted 35–40 days before the tomatoes are planted in the seed bed or transplanted to the open field, and attract a rich natural enemy fauna including *Orius, Chrysopa,* and others.

Research has been carried out on the benefits provided by polycultures in terms of modifying the physical environment of the agroecosystem. For example, strip-cropping of maize in tomato during periods of environmental stress improves fruit formation and yield. Maize planted in double rows between every few rows of tomatoes, 40–45 days before transplanting, has proven to be a good protective crop for tomato. Higher yields were obtained in this intercrop compared to tomato monocultures (Pino 1997).

Legumes play a particularly important role in polyculture systems. Higher yields, better soil fertility, and fewer weeds were found in a maize–*Canavalia* combination (Treto et al. 1997; García 1998, personal communication). Similar results have been obtained in maize–velvet bean polycultures (Guzmán, personal communication 1995).

Empirical observations on farmers' fields have revealed successful intercropping combinations that include legumes, cucurbits, Aliaceae, Solanaceae, and grains such as maize and rice. Polycultural systems play an important role in the development of sustainable agriculture, providing greater yield stability and food security in farming regions.

References

Amador, M.F. and S.R. Gliessman, 1989. An ecological approach to reduce extension impacts through the use of intercropping. *Agroecology.* Ecological Studies 78. University of California, Santa Cruz, pp. 146–159.

Caraza, R., C. Huerres, and C. Pereira. 1996. Sistemas de rotación y asociación de cultivos para primavera verano en organopónicos. *Agricultura Orgánica* 2 (3): 14–16.

Casanova, A. 1995. Experiencia en la producción de hortalizas en condiciones de organopónicos. *Memoria Taller.* Havana: FAO, pp. 68–74.

Casanova, A. and J. R. Savón. 1995. Producción biointensiva de hortalizas. *Revista Agricultura Orgánica* 1 (3):13–16.

Hernández, A. 1998. Evaluación de genotipos de yuca (*Manihot esculenta*) y frijol (*Phaseolus vulgaris*) en un sistema policultural (master's thesis). Havana: ISCAH-CEAS.

Hernández, A., R. Ramos, and J. Sanchez. 1997. Posibilidades de la yuca en asociación con otros cultivos. III Encuentro Nacional de Agricultura Orgánica. Villa Clara, Cuba, pg. 36.

Leihner, D. 1983. Yuca en cultivos asociados. Manejo y Evaluación. Cali, Colombia: CIAT.

Leyva, A. 1993. Las asociaciones y las rotaciones de cultivo in Primer Curso de Agricultura Orgánica. Havana: Agrarian University of Havana.

Mojena, M. 1998. Arreglos espaciales y cultivos asociados en yuca. Modificaciones en algunas variables del agroecosistema y su influencia en los rendimientos totales (dissertation). Havana: Agrarian University of Havana.

Mojena, M., M.P. Bertolí, P. Marrero, and M.D. Ortega. 1996. Asociaciones de cultivos con yuca (*Manihot esculenta* Crantz), una forma de aprovechar el espacio disponible INCA. X Seminario Científico. San José de las Lajas: INCA.

Pino, M. 1997. Informe de etapa. Proyecto 002 00 105. Havana: Liliana Dimitrova Horticultural Research Institute.

Quintero, P.L. 1995. Uso de los policultivos en áreas de producción agrícola. En MINAG-IIHLD-ACAO. I Curso Taller "Sistemas de cultivos múltiples." Havana: (s.n.) pg. 18–21.

Quintero, P.L. 1998. Evaluación de asociaciones de cultivos en la provincia La Habana (technical report). Cooperativa de Producción Agropecuaria Gilberto León.

Quintero, P.L. 1999. Evaluación de algunas asociaciones de cultivos en la Cooperativa Gilberto León de la provincia La Habana (master's thesis). Universidad Agraria de La Habana. Havana: CEAS.

Quintero, P.L., A. Hernandez, N. Pérez and A. Casanova. 1997 Influencia de la asociación boniato–maiz en la regulación biológica del tetuan. *Resumenes del III Encuentro Nacional de Agricultura Orgánica.* Villa Clara, Cuba: Central University of Los Villas.

Treto, E., N. Pérez, O. Fundora, A. Casanova, L. Angarica, and F. Funes. 1997. Algunos resultados del proyecto SANE-Cuba (1995–1997). *Resumenes del III Encuentro Nacional de Agricultura Orgánica.* Villa Clara, Cuba: Central University of Los Villas.

Vandermeer, J. H. 1989. *The Ecology of Intercropping.* Cambridge, MA: Cambridge University Press.

CHAPTER NINE

Mechanization, Animal Traction, and Sustainable Agriculture

Arcadio Ríos, Agricultural Mechanization Research Institute and Félix Ponce, Agrarian University of Havana

The Spanish first introduced the use of cattle as draft animals in Cuba nearly five centuries ago. Until the latter half of the twentieth century, locally adapted breeds of oxen played the predominant role in soil preparation, cultivation, and transportation. Horses, donkeys, and mules were occasionally used for soil preparation as well, though they were more frequently used for transporting coffee and other products in the rugged highlands.

Before the triumph of the revolution in 1959, agricultural mechanization was very limited; most activities were accomplished by manual labor or animal traction. In 1960 there were some 500,000 oxen, 800,000 horses, and 350,000 mules in Cuba. There were also approximately 9,000 low-horsepower tractors (Ríos 1995). From 1959 through 1990, the number of tractors grew rapidly, with a corresponding drop in draft animals (see Table 1). With the shortage of fuel, spare parts, tires, and new machines brought on by the crisis of the 1990s, these trends reversed themselves.

Table 1. Number of tractors and draft animals in Cuba from 1960 to 1997

	1960	1970	1980	1990	1997
Tractors	9,000	52,000	68,000	85,000	73,000
Oxen	500,000	490,000	338,000	163,000	400,000
Draft Horses	800,000	741,000	811,000	235,000	282,000
Mules	30,000	29,000	25,000	30,000	32,000

(Ríos and Aguerreberre 1998)

"Tractorization" and its Consequences

In the past the heavy utilization of draft animals in Cuban agriculture was a function of the lack of motorized vehicles. During the first years of the Revolution there was a massive introduction of tractors as part of a strategy to transform and modernize agriculture. The socialist model of high-tech agricultural production on large-scale state farms promoted "Green Revolution" technology, which went hand-in-hand with mechanized production methods. Between 1970 and 1990, the number of tractors increased tenfold, reaching 85,000 in 1990. This was an increase not only in quantity but also in quality—the average horsepower of the tractors increased from 40HP to 75HP. In this same period, the number of oxen dropped to 163,000 head (Ríos and Aguerreberre 1998).

By 1990, the use of draft animals was limited to smaller cooperatives and private farms. The former Soviet Union supported mechanization by providing credits for importing tractors, combines, other farm machinery, spare parts, and fuel. At the same time, food imports from the former Soviet Union and the Socialist Bloc grew to more than 50 percent of total consumption. As a result, the average yields, economic structure, and social organization of agriculture changed dramatically, as did the eating habits of the Cuban population.

By 1980, in the heat of the Green Revolution, Cuba reached a high level of productivity and scientific and technical development. During this period, the Cuban population grew to twice that of the prerevolutionary period. Meanwhile, rural development was a key focus of the revolutionary government. Bringing all possible social benefits to the countryside meant building new communities with the full gamut of social services, electricity, potable water, education, and so on. The twin results were an urbanization of the countryside and a flight of farm labor into other economic sectors. Agricultural production was gradually increasing through the use of tractors, irrigation equipment, and chemicals, as the characteristics of high-input intensive agriculture became ubiquitous. Unfortunately, this type of agricultural production with its reliance on external inputs and heavy mechanization eventually generated a series of agroecological and soil problems including severe soil compaction, erosion, salinization, and waterlogging.

SOIL COMPACTION: In sugarcane, soil compaction caused a gradual decline in yields and substantially reduced the longevity of sugarcane stands. The compaction was a product of the excessive use of tractors, plows, combines, trailers, spray rigs, and other heavy equipment. Studies in Cuba have shown that the continuous use of tractors for soil preparation is five to eight times more damaging than that of similar tillage operations with draft animals

(Ponce et al. 1996). According to Carrobello and Díaz (1998), there are 2.5 million hectares of land in Cuba with varying degrees of soil compaction.

SOIL EROSION: Excessive tillage operations combined with the elimination of protective cover crops and crop residues for prolonged periods of time led to substantial soil erosion in Cuba. Additionally, large gaps in time between soil preparation and sowing operations give wind and rain increased opportunity to cause soil erosion. Minimum tillage operations under intensive agricultural practices were essentially nonexistent. An estimated 4.2 million hectares have been degraded by erosion in Cuba.

SALINIZATION AND POOR DRAINAGE: Salinization and poor drainage occur over a substantial portion of Cuba's productive areas. Although studies on appropriate technologies for reclaiming saline areas have been carried out, the results have not been implemented on a large scale. Successful large-scale recovery efforts are still needed, as there are more than one million hectares of saline soils to be recovered. Soil drainage efforts have been more successful, particularly in sugarcane areas, but approximately 1.5 million hectares are still affected (Carrobello and Díaz 1998).

All of these problems can be attributed to intensive use of agricultural machinery and inappropriate tillage implements; most damage could have been avoided though appropriate soil conservation practices. By the early 1990s, less damaging traditional and/or alternative cultivation techniques were used only by individual farmers and in the cooperative sector, and were prevalent only in certain crops, such as tobacco.

The fall of the former Soviet Union and other socialist countries produced a dramatic drop in the foreign exchange needed to purchase fuel, machinery, and spare parts. This quickly made the former level of agricultural mechanization unsustainable. As a result there was a dramatic overall decline in the use of mechanized farming strategies.

Maintaining food production at levels sufficient to feed the population became an enormous task. As a result, new policies and agricultural strategies were developed to cope with the loss of affordable inputs. The utilization of draft animals rapidly increased and development of supporting infrastructure followed. Changes included the design and construction of new farming implements, and the substitution of tractors and other nonessential machinery.

Scaling Down: Draft Animals and Soil Conservation

One of the most important measures taken in response to the economic crisis was the transformation of a considerable part of the massive state agricultural farm enterprises into the Basic Units of Cooperative Production (UBPC). These new cooperatives enter into a quota contract with the state enterprise that previously owned them for specific products at fixed prices. Whatever is produced beyond the quota receives a higher price or can be sold in the free market. This has led to increased crop diversity and new crop rotations on farms that previously only produced one or two crops. Additionally, small plots of land were given to families for the production of specific crops such as tobacco, cocoa, and coffee. These crops are very labor intensive and well suited to small family holdings as they are high in value and can be grown in a small area.

The growth of the cooperative and individual farming sectors is significant—they now encompass more than 70 percent of total agricultural land. These smaller units, created by the new structure, have more appropriate conditions for the adoption of low-input production technologies and soil conservation practices including animal traction, use of biological fertilizers, biological pest control, and a general reduction in the use of toxic chemicals. The most significant factor encouraging the adoption of these conservation strategies has been the reemergence of a closer relationship between farmers and their land.

Draft Animals in the New Agricultural Model

A critical element of agriculture during the "Special Period" is the increased use of animal traction. The result is a production system with a reduced use of fuel-driven machinery as well as a decrease in the use of herbicides. This transition was difficult for those farmers who had become accustomed to driving tractors. They were not skilled in working with ox teams, and technicians, administrators, and some farmers viewed the use of animal traction as a regression to the past, while tractors were perceived as a sign of progress.

In 1992 the Ministry of Agriculture (MINAG) and the Ministry of Sugar (MINAZ) established a new program promoting the utilization of draft animals. At this time there was strict rationing of fuel and spare parts for tractors, which had to be purchased with scarce foreign currency. The program encountered numerous problems getting started due to the limited infrastructure for farming with draft animals. The first major barrier was the purchase of oxen for training. The Ministry of Agriculture attempted to purchase large numbers for the agricultural sector, but there simply were not enough to buy. Thus, only a very limited number of male cattle were slaughtered for

Animal traction with an ox team.

consumption, and essentially all the bulls in good physical condition were selected and delivered to cooperative and state farms. The first program in 1991–1992 provided an initial supply of 100,000 bulls. The second round of the program in 1993–1995 provided an additional 100,000 head. The program has built up the number of oxen to 376,000, almost 2.5 times the original number in 1990. Cuban farmers are now training almost 30,000 oxen annually to increase the number of ox teams and to replace those too old to work (Ruiz 1998).

None of the large state farms or even the mechanized cooperatives had the necessary infrastructure to incorporate animal traction into their farming systems: few people could train the animals and ox drivers; pasture and feed production did not exist on site; and at first there were problems of feed transportation. Finally, sufficient veterinary services and technical support were not yet in place for this transition, so keeping the animals in good health was difficult. But slowly mechanisms were put into place to deal with each of these problems.

Table 2 shows that draft animals are especially important in the private sector, which uses 78 percent of the draft animals on only 15 percent of the land. The remaining state sector (which initially had a much higher level of mechanization), the UBPCs, and the Agricultural Production Cooperatives (CPAs) together use only 22 percent of the animals on 85 percent of the land (MINAG 1997, adapted by Ríos and Aguerreberre 1998).

Another important issue regarding the adoption of draft animals is tradition. In certain regions of the country, draft animals were always used on a large scale despite the widespread introduction of tractors. In tobacco production, for example, various agronomic operations have always been done by hand to guarantee high tobacco leaf quality. The use of animals has remained prevalent in these crops, despite the move toward mechanization in other sectors. These farmers became a critical resource for the training program for animals and their drivers.

The level of mechanization in a given area also depended on the overall level of economic development in the region. For example, Havana province has a high level of mechanization and is much more developed than Pinar del Río, where the majority of work was always done with draft animals. The same is true for other provinces where draft animals were predominantly used (Ponce et al. 1996).

Table 2. Distribution of oxen by agricultural sector in 1997

SECTOR	NUMBER OF OXEN	PERCENT OF TOTAL
State Production Units	40,000	10
Agricultural Production Cooperatives (CPAs)	16,000	4
Service and Credit Cooperatives (CCSs) and individual farmers	312,000	78
Basic Units of Cooperative Production (UBPCs)	32,000	8
Total	400,000	100

(MINAG 1997, adapted by Ríos and Aguerreberre 1998)

Development of New Technologies

Conventional soil preparation techniques relied on disc plows and furrow discs pulled by tractors, and on traditional wooden plows pulled by animals. These conventional tillage and cultivation operations led to a loss of soil fertility over time, higher indices of weed infestation, severe soil erosion, and many other problems, some of which are irreversible. A fundamental change in recent years has been the development of a new type of plow by the Agricultural Mechanization Research Institute (IIMA) and the Soil and Agricultural Chemistry Research Institute (IISA). The new "multi-plow" has various versions designed for both tractors and ox teams. The multi-plow was designed for plowing, harrowing, ridging, and tilling. It can also be used for sowing, covering, hilling, and other operations. This plow is completely different from the disc plow or furrow disc because it opens the soil horizontally, but does not invert the topsoil layer. The result is a system that does not mix the different soil layers, as is the case with conventional plows that tend to produce long-term loss of soil fertility. Another advantage of the multi-plow is that it

Cuban-fabricated multi-plow for animal traction, Agricultural Mechanization Research Institute.

helps with weed control, especially of rhizome grasses, which disc plows cut into many pieces, each of which remains viable, resulting in prolific weed growth.

Blacksmith and harness shops were established to manufacture the new implements. The IIMA and other innovators throughout the country are continually improving the new farming implements such as the multi-plow, creating next-generation implements like the "6 in 1," which makes it possible to perform six or more operations making only simple adjustments. Other equipment for animal traction, such as grain and potato sowing devices, multi-tillage equipment, and wide tillage machines and sprayers are gradually being introduced. Table 3 shows that the number of animal traction implements increased 2.34 times between 1990 and 1997, and the number of blacksmith shops 5.6 times. Although the production of new implements exceeds 11,000 units per year, there was only a slight increase in the number of implements per animal due to the large increase of animals over the same period (MINAG 1997).

Table 3. Animal traction implements and blacksmith shops from 1990 to 1997

INDICATOR	1990	1997	RATIO: 1997/1990
Animal traction implements	160,000	375,000	2.34
Number of implements/ox team	2.00	2.08	1.04
Blacksmith shops	500	2,800	5.6

(MINAG 1997, adapted by Ríos and Aguerreberre, 1998)

Education and Training for Animal Traction

The educational level of farmers, fieldworkers, managers, and agricultural technicians has an important effect on the incorporation of agricultural technologies into field practice. In many Third World countries, the technical knowledge and formal education of farmers is very poor, and this decisively affects the potential to incorporate emerging technologies into farming practices (Starkey 1988). In contrast, the educational level of farmers in Cuba is relatively high, and highly qualified personnel work at all levels of management and administration. This made it easier to introduce new technologies and equipment such as tractors and combine harvesters. It also contributed to a preference for mechanized, fuel-driven implements in the post-revolution generation of farmers and agricultural professionals.

At the initial stage of the program to boost the use of animal traction it became evident that a widespread program of technical demonstrations and training would be necessary. This effort included demonstrations of the use of new implements, the selection and promotion of the most appropriate implements by region and crop, and competitions and exhibitions of ox drivers and teams, blacksmiths, and manufacturers of yokes and harnesses. Many farmer-to-farmer exchanges on this topic were organized.

These events cover not only animal traction but also mechanized traction, transportation, irrigation, and so on, to provide a broad knowledge base; activities have engaged thousands of tractor operators, drivers, mechanics and shop workers, farmers, trainers, technicians, and agricultural leaders from most of the production units in all sectors (state, UBPC, CPA, *campesino*) at the municipal, provincial, and national levels. In 1997 alone, a total of 2,344 events were held (1,818 of them were on-farm), with a total of 64,279 participants (MINAG 1997).

Research institutes and the educational network for training qualified workers, technicians, and professional personnel played an important role in the development of mechanization, and now in the assimilation of new technologies using animal traction. Each province has one or more training centers for training and retraining of personnel. Throughout the country, there are numerous Agricultural Polytechnic Institutes (IPAs), or rural vocational high schools, and agricultural universities. The Ministry of Agriculture and the Ministry of Sugar have organized a national network of 19 agricultural research institutes with facilities throughout the island. Additionally, the Ministry of Higher Education (MES) and the Ministry of Science, Technology, and Environment (CITMA) have other research and training institutions. These facilities are adapting and improving animal traction technologies in their regions and training local farmers and professionals in their use.

Conclusion

The recent rediscovery of animal traction farming methods in Cuba is illustrated by the more than two-fold increase in the number of oxen since 1990. But Cuba still has a long way to go to reach its full potential in terms of the number of animals, farm implements, and harnesses, and in the number of trained personnel.

Something that is widely known but rarely taken into account is that mechanization, animal traction, and manual labor produce different impacts on one of our most important resources—the soil. These technologies should not be considered mutually exclusive but rather they should be considered complementary. It is not simply a question of replacing the tractor, but rather the careful selection of alternatives that meet many criteria—always keeping cost efficiency in mind. However, in the case of Cuba other factors have driven policy changes. Because fuel and other farm inputs have become so expensive, current policy favors animal traction to save fuel as well as preserve the health of the soil.

References

Carrobello, C. and R. Díaz. 1998. Agricultura en Cuba. *Revista Bohemia* Vol. 90, No. 17.

MINAG. 1997. *Dictamen de la comisión de mecanización y tracción animal.* II Encuentro Nacional de Mecanización y Tracción Animal. Yaguajay, Cuba.

Ponce, F., R. Torres, and R. Vento. 1996. *Determinación del grado y la intensidad de apisonamiento del suelo por los animales de tracción y los tractores ligeros.* II Congreso Internacional de Tracción Animal. Havana: FAO-IIMA.

Ríos, A. 1995. *Improving Animal Traction Technology in Cuba.* Proceedings of the ATNESA Workshop. Nairobi, Kenya.

Ríos, A. and S. Aguerrebere. 1998. *La tracción animal en Cuba.* Evento Internacional Agroingeniería. Havana.

Ruiz, P. 1998. *La mecanización en el Ministerio de la Agricultura.* Conferencia en el Evento Internacional Agroingeniería. Havana.

Starkey, P. 1988. *Perfected Yet Rejected: Animal-Drawn Wheeled Toolcarriers.* Eschborn, Germany: Vieweg for German Appropriate Technology Exchange.

CHAPTER TEN

Advances in Organic Soil Management

Eolia Treto and Margarita García, National Institute of Agricultural Sciences, Rafael Martínez Viera, National Institute for Fundamental Research on Tropical Agriculture, and José Manuel Febles, Agrarian University of Havana

There are five distinct historical periods in soil management and conservation in Cuba (Table 1). Before 1492, Cuba was covered with virgin soils and forests, and was inhabited by indigenous peoples who coexisted with nature. During four long centuries as a Spanish colony, agriculture was introduced, based mainly on sugarcane and coffee. During this time intense deforestation and burning of crop residues led to soil degradation. This was followed by a relatively brief period of 57 years as a neocolony of the United States, during which time our soils were further degraded by commercial practices.

Table 1. Historic periods and their effect on soil conservation or degradation in Cuba

PERIOD	NO. OF YEARS	CHARACTERISTICS
Precolonial, before 1492	–	Virgin soils, covered with forests, aboriginal communities in equilibrium with nature
Spanish colony (1492–1902)	410	Virgin soils, the start of commercial agriculture, the start of degradation, some conservation, small and medium farms prevail
North American neocolony 1902–1958	56	Rapid degradation with little soil conservation, large plantations with monocrops prevail
Socialist Cuba (before the collapse of the Socialist Bloc), 1961–1989	28	Beginning of soil conservation policy. Large state enterprises with high input practices prevail. Soil degradation continues, though less rapidly
Socialist Cuba (after the collapse of the Socialist Bloc) 1990–2000	10	Rapid increase of soil conservation practices, small and medium properties with low-input practices predominate

In the four centuries that Spain controlled Cuba politically and economically, half the forests disappeared. Then, after only fifty years of US neocolonialism, only 14 percent of the forest area remained (Figure 1). Large, private plantation holdings of thousands of hectares emerged, dedicated only to sugarcane, putting monoculture in the forefront of Cuban agriculture and driving further soil degradation. During this period there were no policy initiatives to counter these trends.

Over the next thirty years the life of the country changed dramatically, as Cuba embarked on its socialist revolution and received economic aid from the Socialist Bloc in Eastern Europe. Although soil degradation continued its advance, the first serious conservation measures were implemented, indiscriminate tree cutting was stopped, and an organized reforestation program was set in motion. This effort reversed the trend toward deforestation, and made some inroads in restoring forest areas (see Figure 1). Various soil research institutes were founded, and the Ministry of Agriculture (MINAG) created the General Directorate of Soils and Fertilizers (DGSF) which engages in soil improvement activities. The vast plantation holdings of the prerevolutionary era were nationalized, becoming large state farms. Here the production model was characterized by heavy machine use, large-scale irrigation, intensive fertilizer and pesticide use, extensive monocultures, and other practices that degrade the soil.

Figure 1. Deforestation in Cuba

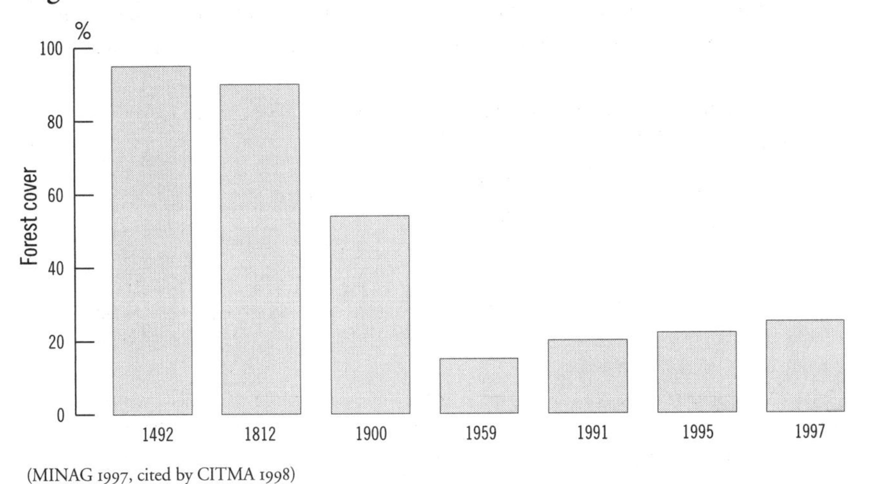

(MINAG 1997, cited by CITMA 1998)

The fifth historical period began in 1990. In spite of the severe economic crisis brought on by the collapse of the Socialist Bloc and intensification of the US embargo, Cuba remained a socialist country. The economic crisis, combined with serious man-made ecological problems, has profoundly changed the fundamental structure and nature of agriculture in Cuba.

Large state enterprises were divided up into medium- and small-sized cooperative farms, which have diversified their production. Massive efforts were undertaken to increase agricultural production in and around cities. A new alternative model of agriculture was developed to reduce imported chemical inputs and dependence on fossil fuels, by using organic fertilizers, animal traction, crop rotation, polycultures, and other soil conservation practices.

Surveys have shown that at least 14 percent of Cuba's land area is affected by erosion and desertification, and that a very large percentage of our soils are affected by ongoing degradation processes (CITMA 1998) (see Table 2). In response the Ministry of Science, Technology, and Environment (CITMA) established new laws and programs for soil conservation and environmental protection, such as Law 81 for Environmental Protection. CITMA also established the National Environmental Program and the National Environmental Strategy, both of which include soil conservation and recovery as priority areas. Recently CITMA established the National Action Program Against Desertification and Drought. This program has begun work in each watershed of the country, using a multidisciplinary and integrated methodology of watershed management. Another recent national program that includes soil conservation is the Integrated Development Program for Mountain Areas.

Table 2. Area affected by soil degradation

DEGRADATION PROCESS	TOTAL AREA AFFECTED (MILLIONS OF HECTARES)	% OF AGRICULTURAL AREA AFFECTED
Salinization	1	14.9
Erosion (medium to high)	2.9	43.3
Poor drainage	2.7	40.3
Poor internal drainage	1.8	26.9
Low fertility	3.0	44.8
Compaction	1.6	23.9
Acidity	1.7	24.8
Very low organic matter content	2.1	31.8
Poor water retention	2.5	37.3
Rocky	0.8	11.9

(MINAG 1996, cited by CITMA 1998)

The financial and ecological problems affecting our country have led us to accelerate the implementation of alternatives to maintain and improve soil fertility. By combining traditional methods with modern science, Cuba has implemented major alternative programs such as the use of residues from the sugarcane industry, biofertilizers based on microorganisms, large-scale production and utilization of earthworm humus, composting, green manures, and various soil amendments. Projects exploring minimum tillage for soil conservation have been revitalized. This chapter summarizes Cuban research on the use of these alternatives, and their introduction in commercial-scale production.

Recycling Sugarcane and Other Residues as Organic Fertilizers

Several by-products of sugarcane harvesting, and sugar milling and processing, are currently being used as organic soil amendments on a very large scale. One of the harvest residues, which has been widely used and studied in Cuba as an organic fertilizer, is *cachaza* (filter cake mud). Various reports have been published on its use as a fertilizer in sugarcane and other crops, dating as far back as 1917. *Cachaza* is made up of tiny fibers filtered from the cane juice before it is reduced. It is highly valued as a fertilizer and has high percentages of phosphorus, calcium, and nitrogen, and to a lesser extent, potassium. It generally contains more than 50 percent organic matter.

The availability of *cachaza* depends on the amount of ground sugarcane processed, and constitutes three to four percent of the total harvest weight. In sugarcane cultivation, it can completely replace chemical fertilizers for three years in sandy soils, and for approximately five years in soils with higher clay content. This is achieved with applications of 120 to 160 t/ha and 180 and 240 t/ha, respectively (see Table 3). The dosage can be reduced by ⅓ to ¼ if localized applications are made. Studies have shown yield increases of approximately 10–20 t/ha/year higher than with chemical fertilizer (Arzola et al. 1990).

When applied to pineapples, *cachaza* gave better results than chicken manure, peat, or bagasse. Good results were attained by applying 80 t/ha, which supplied all the phosphate, 90 percent of the nitrogen and 40 percent of the potassium needs of the crop (Treto et al. 1992). Over a span of four years, 70 percent of the nutritional requirements of young citrus trees were met, doubling production without chemical fertilization. Furthermore, good results were obtained in root stock production of sour orange trees *(Citrus aurantium)* in nursery conditions, using a mixture of 50 percent red ferrolitic soil and 50 percent *cachaza,* with no need to apply any chemical fertilizer. The trees grew faster on average than those in soil alone (30 days trial). Satisfactory results have been obtained in other crops such as coffee, plantain, garden vegetables, rice, and vegetables growing in intensive raised beds or organoponics (Peña et al. 1995).

Table 3. Potential of *cachaza* to replace chemical fertilizers for a harvest of eight million tons of sugar (projection)

NUTRIENTS	TONS *CACHAZA*	CHEMICAL FERTILIZER	TONS, CHEMICAL FERTILIZER
N	19,000	Urea	41,300
P (P_2O_5)	23,250	Simple super-phosphate	119,230
K (K_2O)	5,200	KCl	8,750
Ca (CaO)	44,250	$CaCO_3$	83,500

(Arzola et al. 1990)

There have been several proposals regarding the use of diverse residues from sugarcane warehousing facilities. Table 4 shows the average composition of residues from the La Paulina sugarcane collection center at the Abraham Lincoln Agro-Industrial Complex (CAI) in Havana province (Paneque et al. 1986). In a normal harvest, the residues left at a sugarcane collection center are enough to fertilize 160 ha at a rate of 4.3 t/ha, which would contribute with 13 kg/ha of N, 63 kg/ha of P_20_5 and 112 kg/ha of K_20.

Table 4. Chemical composition of crop residues at the La Paulina sugarcane collection center (based on dry weight)

pH	8.3	Ca, %	24.6
M.O., %	15	Mg,%	0.9
N, %	0.33	$CaCO_3$,%	27
P_2O_5 ,%	1.63	Soluble Na, ppm	94
K_2O, %	2.89	Moisture, %	10

(Paneque et al. 1986)

The liquid residues from the sugar processing industry are another important source of plant nutrients. In Cuba approximately 47 million m^3/year are produced each harvest, polluting rivers, seas, and reservoirs when they are not used for irrigation and fertilization. Paneque and Martínez (1992) conducted several trials at the National Institute of Agricultural Sciences (INCA) and concluded that it is possible to use the liquid residues from sugar processing, alcohol distilleries, and torula yeast plants for sugarcane irrigation and fertilization. Table 5 shows the chemical composition of these residues.

Table 5. Chemical composition (kg/m^3) of liquid residues from different industries based on sugar

Residues	N	P_2O_5	K_2O	CaO	MgO	MO
Distilleries	0.37	0.46	6.87	3.08	1.21	57.4
Torula yeast plants	0.22	0.97	1.72	0.27	0.36	2.4
Sugar mills	0.02	0.02	0.09	–	–	1.2

(Paneque et al. 1991; Paneque and Martínez 1992)

Liquid residues from alcohol distilleries are particularly rich in nutrients. Paneque and Martínez (1992) found they could be used in irrigation combined with the liquid from sugarcane processing in a 1:7 ratio. If three such irrigations of 500 m^3/ha each are applied, the nutrient load would be 56 kg of N, 69 kg of P_20_5, 1,030 kg of K_20, and 8.6 tons of organic matter, which is roughly equivalent to an application of 53 t/ha of *cachaza.* This fertilization would last for four sugarcane harvests. Based on these and other results, a manual for the application of liquid residues from the sugar industry for irrigation and sugarcane fertilization has been published (Paneque 1999).

Liquid residues from the citrus industry are also being studied for their potential as fertilizers. If 500 m^3/ha of liquid residues from the citrus industry are applied, 85 kg of N, 25 kg of P_20_5, 125 kg of K_20, and 5.5 tons of organic matter are provided. One citrus processing plant generates 2,500 m^3/day of liquid residues, working 300 days/year. Thus its contribution would be 750,000 m^3 with a potential equivalent to 622 tons of ammonium sulfate, 192 tons of simple super-phosphate, and 312 tons of potassium chloride (Paneque and Grass 1992).

Biofertilizers

Cuba has been studying the use of naturally occurring microorganisms for agricultural purposes for many years. A number of these organisms have shown excellent abilities to improve soil fertility and aid in nutrient and water uptake by the root system. Inoculating plants, seeds, or soils with microorganisms is a method designed to dramatically increase the population of those beneficial microorganisms that help plants to grow.

BACTERIA

Commercial inoculates of strains of *Rhizobium japonicum* have been used for a long time in many countries to stimulate soybean production. Cuba has achieved good results with *Rhizobium,* supplying from 80 to 100 percent of the nitrogen requirements of soybeans. Cuba has national strain selection and production programs for *Rhizobia* inoculation of legume seeds. The development of strain 3412 (ICA 8001), which is effective in numerous commercial Cuban and/or introduced varieties, is a testament to this work (Pijeira and Treto 1983; Pijeira et al. 1988). Other biological fertilizers based on *Rhizobia* strains have been used successfully in beans (Hernández et al. 1998), cowpeas (Hernández et al. 1994), peanuts, and forage legumes (López 1994), reducing chemical nitrogen fertilization by 70 to 100 percent.

Commercial preparations based on the free-living, nitrogen-fixing *Azotobacter chroococcum* bacteria have been extensively used in Cuba. Research has shown that they have a number of beneficial effects on many crops. This bacterium is found naturally in most Cuban soils, but in relatively small numbers (1,000–10,000 cells/g of soil). At these population densities, the beneficial effects of the bacteria are very small. To artificially boost the beneficial effects we must apply commercial preparations to reach populations of approximately 100 million cells/g of soil (Martínez Viera and Dibut 1996a). The Cuban strains currently in use can supply up to 50 percent of the nitrogen requirements of the crop through biological nitrogen fixation. Several biologically active substances (auxins, cytoquinines, giberelins, amino acids, and vitamins) are synthesized by *Azotobacter,* stimulating the growth and yield of economically important crops (see Tables 7, 8, and 9) (Dibut 1998). The combination of these substances, which are assimilated through the roots, allows each substance to act at the moment when it is required by the plant. Thus, some of them promote the development of roots or of the whole plant, others increase flowering or reduce flower losses, while others stimulate fruit development. These effects allow for the more rapid development of more vigorous plants with higher yields.

The results of *Azotobacter* application are offered here as an example of the beneficial effects of biofertilizers in Cuban agriculture. For example, by inoculating tomatoes, an increase in germination was obtained, and seedling survival improved by 30 to 40 percent, yielding a greater number of viable plants per kg of seed, while reducing the area needed to produce seedlings for transplant. The active substances produced by the bacteria also accelerated seedling development. Seedlings were on average 30 percent taller, had 20 percent more leaves, 40 percent greater stem diameter, and 52 percent more biomass. These applications have shortened the period between sowing and transplanting, as plants could be transplanted 7 to 10 days earlier. An additional benefit was savings of water, labor, and plant protection costs, as the whole cycle was shorter (Martínez Viera and Dibut 1996a).

Once transplanted in the field, fruit production came earlier, and the number of fruits per plant was 35 percent greater in-season, and 60 percent greater when grown out of season. This produced an overall average yield increase of 25 percent for in-season production, and a 40 percent increase when grown out of season. This is especially important for maintaining tomato production year-round. Fruit quality was significantly better as well. Commercially speaking, from 80 to 85 percent of the fruit was of high quality, compared to 60 to 70 percent of nontreated fruits. These results were obtained while eliminating 40 percent of the nitrogen fertilizer normally used in commercial production.

In the case of cassava and sweet potato, in addition to nitrogen fixation, the active substances synthesized by the bacteria stimulate photosynthesis and reduce respiration of the crop plant. This favors accumulation of photosynthates, which act as reserves for tuber and root formation. Table 6 shows the yields obtained with the application of *A. chroococcum* on two cassava and two sweet potato varieties (Martínez Viera 1997). In Table 6, one can observe that *Azotobacter* application not only compensates for 50 percent of the recommended nitrogen in these crops, but also increases yields, presumably via the stimulation and regulation of photosynthesis and respiration.

Table 6. Yield effect (t/ha) of application of *A. chroococcum* with different doses of nitrogenous fertilizer on cassava and sweet potato

VARIANT	CASSAVA VARIETIES		SWEET POTATO VARIETIES	
	CMC-90	Señorita	Yabú-8	CEMSA-78 354
100% N	31.5	32.9	25.0	30.0
50% N	20.0	20.6	18.7	20.6
50% N+ *Azotobacter*	32.6	35.4	30.6	34.3

(Martínez Viera 1997)

In plantain, this biofertilizer improves various phenological indices and overall productivity, as long as recommended techniques are followed. Data presented in Table 7 demonstrates that 30 percent of nitrogen fertilizer can be eliminated without yield reductions (Dibut et al. 1996). When the full dose of nitrogen fertilizer is applied, there apparently is no nitrogen fixation, since bacteria absorb the abundant nitrogen and save energy that would normally be used for fixation (exacting a high biological energy cost). At this dose combination, yield increases are due to the action of the other active substances. The experimental use of *Azotobacter* on sugarcane has produced good results as well, but research is still being conducted on the best commercial-scale methods to employ.

Under tropical conditions nitrogen is also fixed in the phylosphere, the zone of contact between the atmosphere and the leaf. The microorganisms that live on the leaves, among them *Azotobacter* bacteria, absorb water and dissolved gases from the atmosphere, as well as nutrients exuded from the green leaves. The foliage functions as a substrate, water trap, and nutrient production center, creating a medium for microbial growth. The plants are then capable of absorbing these nitrogenous compounds, and distribute them towards the rapidly growing parts. This ability to concentrate atmospheric substances gives the leaves an important job in agricultural ecosystems (Bhat et al. 1971).

Table 7. Yield effect (t/ha) of *A. chroococcum* in plantain and banana with different doses of nitrogenous fertilizer

	BANANA (GIANT CAVENDISH)	PLANTAIN (CEMSA)
100% N	42.2	23.8
100% N + *Azotobacter*	44.8	27.7
70% N	38.0	20.5
70% N + *Azotobacter*	41.8	23.1
50% N	34.5	15.1
50% N + *Azotobacter*	36.9	18.1

(Dibut et al. 1996)

The role of the phylosphere was seriously considered in Cuban studies of methods for the use of *A. chroococcum* as a biofertilizer. The research has confirmed that foliar applications can produce dramatic increases in yields. In large citrus plantations, aerial application of *Azotobacter* has allowed for dramatic reductions in nitrogenous fertilizer use, as shown in Table 8 (Martínez Viera and Dibut 1996b).

Table 8. Yield effect (t/ha) of foliar application of *A. chroococcum* to citrus crops (*Citrus grandis* and *Citrus sinensis*)

TREATMENT ON GRAPEFRUIT	YIELD (T/HA)	TREATMENT ON ORANGE	YIELD (T/HA)
50% N + *Azotobacter*	73.0	50% N + *Azotobacter*	48.0
100% N	66.5	100% N	36.3
50% N	57.9	50% N	27.6

(Martínez Viera and Dibut 1996b)

Another useful microorganism in plant nutrition is *Azospirillum brasilense*. This bacterium has the special property of being able to form symbiotic relationships with the roots of cereals much like the relationship between *Rhizobia* and legumes. When *Azospirillum* is associated with certain cereal grains it can fix atmospheric nitrogen and produce plant growth hormones. In Cuba, studies to determine the effect of inoculations on rice and sugarcane were carried out by Velazco et al. and Roldós et al. in 1992. Results of these studies showed that up to 50 percent of the nitrogen requirements of sugarcane, and from 25 to 50 percent of those of rice were met (Velazco et al. 1992). While future research and development of *Azospirillum brasilense* may yield commercial applications, use of this bacterium is still limited.

Mycorrhizae

Enhancing the density of mycorrhizal fungi (MVA) in the soil benefits plant nutrition, because they function as an extension of the plant root system, increasing water and nutrient absorption capacity. Early research on these phenomena came out of forest ecology, where massive underground networks of mycorrhizae fungus can connect thousands of individual trees. In the early 1990s, Cuban forest ecologists were called upon to research their potential for use in agricultural ecosystems. Good results have been obtained in Cuba when citrus trees, coffee, passion fruit, mango, pineapple, and tobacco seeds were inoculated with mychorrizal fungi. During the 1990s, research was done on these crops by Gárciga et al. (1992), by Herrera (1994) at the Institute of Ecology, by Fernández et al. (1997) at the National Institute of Agricultural Sciences, by Cueto and Sánchez (1994) at the Citrus and Fruit Research Institute (IICF), and by Ruiz (1993) at the National Research Institute for Tropical Roots and Tubers (INIVIT).

In plantain nurseries, young in-vitro seedlings are inoculated with *Glomus mosseae* early in their development. Inoculated seedlings grow bigger and faster, increasing biomass by as much as 70 percent over control groups, even when irrigation is cut by 50 percent (Noval et al. 1997). In-vitro pineapple seedlings inoculated with MVA showed biomass increases of 50 to 100 percent, reducing the stage of adaptation for transplant by 15 to 30 days (Noval et al. 1995, 1997).

Bustamante et al. (1992) conducted experiments in applying *Glomus fasciculatum* in coffee nurseries using four typical soil types. They mixed the soils with earthworm castings in a 5:1 ratio. In one soil type seedling height and foliar area increased by 20 and 25 percent respectively; in another the increases were 12 percent and 51 percent; and in another 7 and 25 percent. No differences were found in ferrolitic soils. Fernández et al. (1997) tested five rice and wheat strains with a variety of mycorrhizae. *Glomus mosseae* and *G. manihotis* excelled, increasing yields by 20 to 45 percent in rice on a commercial scale.

Since 1992, small-scale commercial production of biofertilizers, based on bacteria and mycorrhizal fungi, has been on the rise across the country. Commercial formulations based on small-scale production methods developed by the National Agriculture Science Institute (INCA) are sold under the *Azofert, Ecomic,* and *Rhizofert* brand names (INCA 1997). These products have been successfully used in Cuban agriculture and in other countries such as Colombia and Bolivia. They have been used in coffee, citrus, and other fruits, for tomato and other vegetable seedlings, in the adaptation stage of in-vitro seedlings of pineapple, banana, and sugarcane, and in directly-sown crops such as rice, maize, tomatoes, soybean, and sorghum. In direct sowing, biological fertilizers like *Ecomic* employ the novel technique of seed coating (pelleting), also developed by INCA (Fernández et al. 1997).

Phosphorus-Solubilizing Microorganisms

Important microorganisms that specifically help improve the phosphorous nutrition of plants are the so-called phosphorus-solubilizing agents. Dr. Angélica Martínez and her colleagues at the Soil Research Institute (IIS) have studied them in Cuba. The results of their initial research were reported at the first International Biofertilizer Workshop, held in Havana in 1992. Typically, Cuban soils are rich in phosphorous, most of which is bound to soil particles and thus unavailable to crop plants. Using organisms as soil inoculants under the trademark *Fosforina*, farmers are able to release some of the phosphorus tied up in the soil, making partially or totally available the phosphorus needed by crops such as tomatoes, cucumbers, tobacco, sugarcane, and citrus.

Soil type is a determining factor in the impact of *Fosforina* application. In the case of tomatoes, *Fosforina* reduced the need for chemical phosphorous by 75 percent in one soil type but only by 50 percent in another. In a third soil type it covered 75 percent of fertilizer recommendations, while in another it made 100 percent of the phosphorous needed available, while producing yield increases of from 5 to 38 percent.

The application is done by soaking seedlings in nondiluted *Fosforina* for 5 minutes (8 liters/ha) before transplanting, or by spraying the soil with a diluted solution (1:10) at a rate of 20 liters/ha. In tobacco seedlings, increases in biomass of from 30 to 50 percent have been reported—with only 50 percent of normal phosphorus fertilization. This is the current recommendation for Burley 32 and Carrojo varieties sown in certain soil types. Similarly, sugarcane (variety Ja-60-5) showed increased germination with 38 percent higher yields, and increased phosphorus content in juice (46 percent).

Biofertilizer Mixtures

Promising results have been obtained with combinations of various biofertilizers, such as *Azotobacter*, MVA, and *Fosforina*. Nardo et al. (1992) at INCA showed that combining *Fosforina*, MVA, and growth promoters reduced the nursery time for Cleopatra mandarin orange (*Citrus reticulata*) seedlings. At INIVIT Ruiz (1993) obtained a 30 percent yield increase in cassava using a combination of *Azotobacter*, MVA, and *Fosforina*, with a tuber yield of 28t/ha. Bustamante et al. (1992) reported good results from inoculations of coffee plants in nurseries with the same combination. Rivera et al. (1997) had positive results combining MVA and rhizospheric bacteria. Corbera (1998) found that *Bradyrhizobium japonicum* and MVA inoculations in soybean were effective in achieving comparable yields without chemical fertilization.

All in all, microbial fertilizers have played an important role in Cuban agriculture since the onset of the economic crisis. Table 9 shows the most widely used biofertilizers in Cuba and their potential to reduce the need for chemical fertilizers.

Table 9. Principal uses of biofertilizers in Cuba

BIOFERTILIZER	CROPS	SUBSTITUTION ACHIEVED
Rhizobium	Beans, peanuts, and cowpeas	75–80% of the N fertilizer
Bradyrhizobium	Soybeans and forage legumes	80% of the N fertilizer
Azotobacter	Vegetables, cassava, sweet potato, maize, rice	15–50% of the N fertilizer
Azospirillum	Rice	25% of the N fertilizer
Phosphorous-solubilizing bacteria	Vegetables, cassava, sweet potato, citrus fruits, coffee nurseries	50–100% of the P fertilizer
Mycorrhizae	Coffee nurseries	30% of the N and K fertilizers

(Martínez Viera and Hernández 1995)

Manure, Compost, "Bioearth," and Earthworm Humus

The application of animal manures is a time-honored and universal technique used to improve soil fertility and structure. Typical application rates in Cuba range from 25 to 80 tons per hectare. Beyond providing the principal nutrients needed by crop plants (N-P-K), it also restores many of the micronutrients lost during harvest, enriching the soil with the organic matter needed to build a healthy soil ecology.

In Cuba the primary source is cattle manure collected at dairy farms. Once a source of groundwater pollution, this manure is now used in a variety of crops. Sometimes the manure is aged and tested for weed seeds and plant pathogens; at other times it is processed through hot composting, vermiculture, and/or is mixed with other mineral amendments and microbiological inoculants. Crespo and Gutíerrez (1992) have worked on such mixtures, finding, for example, increased effectiveness from mixing manure with zeolite (a mineral soil activator) at a ratio of 8:1 and adding *Fosforina*. This boosts the available nitrogen while doubling phosphorus solubility.

Sheep and poultry manure also give good results. The latter is especially used on acidic soils, as lime is used on poultry farms to aid in pest and disease control, and passes into the manure making it useful for "liming" acidic soils. Caution must be used with chicken manure, especially in crops that are sensitive to high concentrations of calcium and to high pH.

Another source of quality organic fertilizer is compost made from post-harvest wastes and crop residues, grasses, manure, and other organic wastes. In Cuba, compost is often inoculated with microorganisms to increase the

Compost production at an integrated agricultural farm–livestock ranch.

decomposition rate and to enhance the quality of the final product. This type of compost is commonly called "bioearth." Bioearth was developed at Las Villas University (Mayea 1994) and initial field results were obtained at the Nueva Paz Mixed Food Crop Enterprise in Havana Province.

Bioearth is produced by inoculating a mixture of organic materials with *Azospirillum orizae* (a bacterium), *Trichoderma* sp. (a fungus), *Saccharomyces cerevisiae* (brewer's yeast) and *Bacillus nato* (a bacterium). Bioearth is a very high-quality organic fertilizer that gives good results in much smaller doses than manures. Application rates of as little as 6 to 7 t/ha have produced excellent results in vegetable crops, greatly reducing transportation and application costs. It has also been shown to control plant pathogens in the compost itself, as well as in the field.

Another high-quality fertilizer commonly used is worm castings or worm humus. In Cuba, researchers from the Soil Research Institute have successfully promoted vermiculture on a national scale (Ramón-Cuevas et al. 1996; González et al. 1996). According to a manual prepared by the Provincial Delegation of MINAG (1993), organic wastes (mostly manures, and *cachaza)* are used as raw materials and are transformed by earthworms into humus. As the organic material passes through the digestive tract it is converted into a highly mineralized and bacteria-rich material that can be easily handled, transported, stored, and applied as a soil amendment. Production is carried out both on a small scale using kitchen and other wastes, and at a commercial scale. The worm species used is the California Red (*Eisenia foetida*).

The California Red is used because it tolerates the very high population densities (10,000–50,000 worms/m^2) needed to promote rapid decomposition. It is a very prolific worm, which begins reproduction at just three months of age, mating thereafter every 10 days. Each egg capsule yields an average of 2 to 20 live worms. As a result it is characterized by extremely fast reproduction, especially under high ambient temperatures. Additionally, it is a hearty animal that can resist variations in temperature, pH, and humidity, and prospers in a variety of substrates.

In commercial-scale production, the California Red can produce 2,500 to 3,500 m^3 of humus from 9,000 m^3 of organic material, on a single hectare of land. This is under subtropical conditions where worms can be reared in three harvests per year. Because of the quality of the humus, good results have been attained in crop production with relatively small amounts of humus (2.5 to 5 t/ha). For example, phosphorus requirements of potatoes were met in this fashion. Good results were attained with corn by applying 2.5 t/ha of humus, combined with 100kg/ha of nitrogen and 60 kg of P_2O_5 producing corn yields of 12 t/ha. The utilization of worm humus in various crops and dosages, and its potential for replacing chemical fertilizers, are shown in Table 10.

Raised beds for production of worm-castings. "La Renée" experimental station, Quivicán, Havana.

Table 10. Use of worm castings in different crops in Cuba

CROP	DOSAGE, T/HA	EQUIVALENT CHEMICAL FERTILIZER (%)
Potato	5	25–50
Tobacco	4	65 (phosphorus and potassium)
Plantain	10	50
Tomato	4	25–50
Garlic	4	100 (nitrogen)
Onion	4	50–75
Pepper	4	25
Sweet potato	4	25

(Gandarilla et al. 1995)

In peak years production reached 78,000 tons of worm humus and 701,000 tons of compost (see Figure 2). The application of organic fertilizers to non-sugar crops doubled between 1984 and 1990 (see Figure 3). Then there were production shortages at the beginning of the crisis, though recovery is well underway.

Figure 2. Worm humus and compost production in Cuba

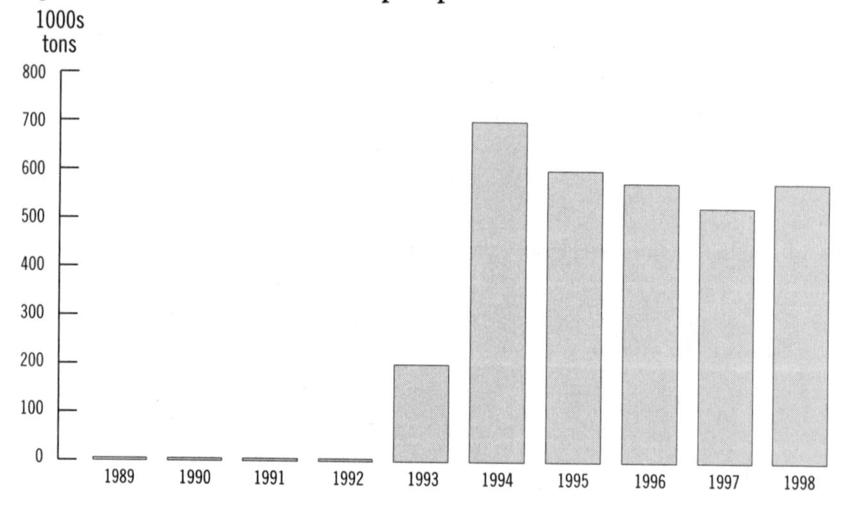

Figure 3. Total application of organic fertilizers in non-sugar crops in Cuba

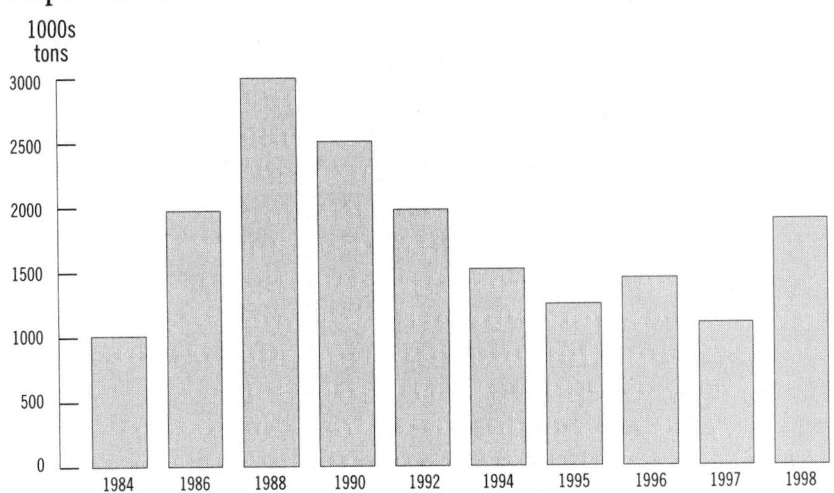

Green Manures

In spite of all the advances in the production and application of worm humus, bioearth, compost, *cachaza*, and biofertilizers, we continue to have difficulties in substituting for all the chemical fertilizers. Often not enough raw materials are available when and where they are most needed. Moreover, taking a macro view, there simply are not enough of these materials to provide soil amendments for all agricultural production at a national level. Even if there were, transportation of such large quantities of material is costly and logistically complicated. Alternatives for soil improvement must be biologically, socially, and economically appropriate to the living standards of farmers, such that off-farm inputs, even organic in nature, cannot fulfill all of the needs. On-farm solutions will have to play an increasing role; thus green manures are extremely important because they open up the possibility of producing large quantities of organic matter *in situ*. Thus they constitute a good strategy for soil management and conservation programs.

At the beginning of the twentieth century many reports on the use of green manures were published by the Santiago de Las Vegas Agricultural Station (1907, 1909, and 1917). A group of tropical legumes, which adapted well to Cuban soils, were recommended. These included jack bean (*Canavalia ensiformis*), velvet bean (*Mucuna* sp.), crotolaria (*Crotalaria* sp.), cowpea (*Vigna* sp.), pigeon pea (*Cajanus cajan*), soybean, and others. These legumes produce large quantities of biomass, fix nitrogen, and recycle considerable amounts of phosphorus and potassium. Some are used for animal feeds, and they also offer improved soil conservation and weed control.

The use of green manures disappeared in most of Cuban agriculture with the introduction of synthetic fertilizers, mechanized and chemical weed control, and the reduction in use of draft animals. However, some small independent farmers held onto this practice, and when researchers began to investigate these alternatives again in the 1980s, they found that farmers in certain regions continued to use green manures with great success. Peña et al. (1988) reported the common practice of Cuban peasants in Banao, a municipality of Sancti Spíritus province, to use velvet beans as a green manure in rotation with onions. Thus Cancio et al. (1990) recommended minimum tillage and green manures in onions.

Green manures are not limited to nitrogen-fixing legumes, but can include plants that have other important characteristics. According to Peña et al. (1986), under the highland conditions of Escambray, sorghum is used to improve and protect eroded soils. By 15 days after germination, this species covered 100 percent of the area and recycled 133, 25, and 189 kg of nitrogen, P_2O_5, and K_2O, respectively. Cancio et al. (1989) tested three legumes (*Mucuna* sp., *Lablab purpureus,* and *Canavalia ensiformis*) in slightly eroded soils and determined that forage sorghum outperformed legumes in boosting tobacco yields.

In recent years the use of green manures in large-scale commercial production has been most extensive in rice. Cabello et al. (1989) at the Central Rice Research Station concluded that it was possible to use sesbania (*Sesbania rostrata*) sown in the spring and incorporated into the soil after 45 days, as fresh green manure for rice. This produces the 60 to 80 kg/ha of nitrogen needed to yield up to 6 t/ha of rice. There are now sesbania seed multiplication programs at every Rice Agro-Industrial Enterprise (CAI) in Cuba, which guarantees an important component of the nitrogen required by this crop.

In 1992, INCA began an intensive program to select the most promising green manure species and their optimum sowing dates. They are examining the potential to substitute organic means for chemical nitrogen fertilizers, by incorporating suitable species in rotations and/or associations with crops such as squash, potatoes, taro (*Xanthosoma sagittifolium*), maize, and coffee (García 1998). According to this study, the tropical legumes and other plants tested as green manures perform better during the rainy season in Cuba, compared to the dry season. In the former, they contribute a greater amount of biomass and nutrients to the soil: 10 to 67 t/ha of fresh biomass, which is equivalent to 2 to 22 t/ha of dry biomass; 67 to 255 kg/ha of N; 7 to 22 kg/ha of P; and 36 to 211 kg/ha of K (see Table 11). During the dry season, they compete with vegetable crops grown in the dry, cooler winter months. Additionally, many of these plants have difficulties in the winter due to their short-day photoperiods. They begin flowering before the desired vegetative growth has been achieved, limiting their value as green manures.

Table 11. Biomass and nutrients provided by green manures in the rainy season in the province of Havana, 1992 and 1993

SPECIES	WET WEIGHT t/ha	DRY WEIGHT t/ha	N kg/ha	P kg/ha	K kg/ha
Crotalaria	63	7.1	255	21	92
Jack bean	35	3.4	153	11	44
Mucuna aterrima	10	3.2	149	8	44
Sesbania rostrata	29	4.4	141	11	101
Lablab purpureus	28	2.9	121	10	52
Sorghum	67	11.0	188	22	211
Crotalaria	38	2.5	92	9	65
Cowpea	18	2.7	77	8	52
Mung bean	11	3.0	67	8	56
Velvet bean	33	2.6	121	7	38
Phaseolus helvolus	18	2.1	79	9	36
Pigeon pea	54	4.5	135	13	67
Range of variation	10–67	2–11	67–255	7–22	36–211

(García and Treto 1997)

The plants which were found to produce the most biomass and nutrients, and were best adapted to Cuban soils and cropping conditions, were jack bean, crotalaria, velvet bean, cowpea, mung bean, *Sesbania*, sorghum, lablab bean, and *Lupinus albus.*

When planted in rotation with squash, the best species were crotalaria and cowpea (*V. unguiculata*), which produced greater yields than with chemical fertilization. Yields in this crop increased from 4 t/ha without green manure to 10 t/ha using crotalaria, and up to 6 t/ha with cowpea. To reach squash yields of over 14 t/ha it is necessary to add only an additional 60 kg/ha of nitrogen, representing a considerable reduction of the previous dosage recommended for this crop, which was 140 kg/ha of N.

Jack bean is the best species to use in rotation with potatoes (in-season), and substitutes for up to 50 percent of nitrogen requirements (80 kg/ha). The highest potato yields were obtained by combining jack beans with 40 kg/ha of chemical nitrogen, substituting for up to 75 percent of fertilizer.

Intercropping taro with edible legumes such as cowpea and soybean increased total yields by 50 percent (4t/ha) compared to monoculture. Intercropped legumes rapidly covered the space between taro rows, controlling weeds and conserving soil moisture. In the rainy season, maize showed promising results when intercropped with each of the following four legumes: *Sesbania*, velvet bean, jack bean, and crotalaria (Treto et al. 1994). In the dry season, maize gave good results when sown with jack bean, devil bean (*Phaseolus helvolus*), and *Lupinus.*

Canavalia is the most resistant legume under the adverse dry season conditions of highland coffee areas, contributing 4.3 t of dry matter/ha (19 t/ha green matter) and 182, 13, and 76 kg/ha nitrogen, phosphorus, and potassium, respectively.

Economic studies on the use of green manures in agriculture revealed savings ranging from 623 to 1,503 Cuban pesos/ha, according to the crop and species used. The increase in profits came from the higher crop yields obtained with these systems and the savings due to reduced use of chemical fertilizers (García 1998). Overall, green manures can substitute for from 50 to 70 percent of the nitrogen requirements of various crops in Cuba, supplying from 51 to 110 kg/ha of N (see Table 12).

Table 12. Examples of nitrogen fertilizer savings from using green manures in Cuba

PERIOD	CROPS	OPTIMUM N DOSAGE kg/ha	GREEN MANURE	MAXIMUM DOSAGE	SUBSTITUTION DOSAGE kg/ha	SUBSTITUTION DOSAGE %	INCREASE IN YIELD t/ha
1986–1989	Rice	100–150	*Sesbania*	75	70–85	50–75	1
1992–1997	Potato	164	*Canavalia*	53	111	68	2
1992–1997	Potato	131	*Canavalia*	67	64	49	4
1992–1997	Squash	110	*Crotalaria*	59	51	46	3,5
1992–1997	Squash	110	Cowpea	35	75	68	–
Range					51–111	46–75	1–4

(Cabello 1989; García 1998)

Green manures are not only a substitute for chemical fertilizers, but also improve ecological soil properties. They also protect soil from the direct effects of the sun, and from the effects of torrential rains and high-velocity winds. They aid in maintaining soil moisture and cool temperatures, and thus they support the growth of beneficial soil organisms. Finally, they lead to measurable improvements in the physical structure of the soil, as shown in Table 13.

Table 13. Impact of green manures on physical properties of the soil in samples from 0–30 cm deep, 60 days after being incorporated

TREATMENTS	HH, %	NH, %	AD, G/CM^3	TD, G/CM^3	TP, %	AP, %
Natural vegetation	4.9 c	29.5 b	1.1 a	2.64 a	58.1 b	24.5 b
Vigna unguiculata	5.1 b	31.6 a	1.0 ab	2.61 b	59.9 ab	33.7 a
Crotalaria juncea	5.3 a	32.4 a	0.9 b	2.62 b	62.5 a	34.7 a

Note: HH, hygroscopic humidity; NH, natural humidity; AD, apparent density; TD, true density; TP, total porosity; AP, aeration pores

(García 1998)

The results of many of these studies have been extended to farmers throughout the country. The maize–lablab (*Lablab purpureus*) association was introduced with satisfactory results, the latter serving as animal forage and green manure, at the "28th of September" Agricultural Production Cooperative (CPA) in Batabanó municipality (SANE 1999). The 28th of September is an Agroecological Lighthouse of the SANE (Sustainable Agriculture Networking and Extension) program of UNDP, run by the Organic Farming Group (GAO) of the Cuban Association of Agricultural and Forestry Technicians (ACTAF).

At the Jorge Dimitrov CPA, another Agroecological Lighthouse, in San Antonio municipality, *Canavalia ensiformis* intercropped with corn has been very successful and is being tried with other crops. At the Gilberto León CPA, in San Antonio municipality, various species are intercropped with plantains, and favorable results have been obtained with jack bean, crotalaria, and velvet bean.

In spite of their tremendous potential to solve multiple problems of soil conservation and fertility, the use of green manures is still not the norm in Cuba. For a large-scale expansion in green manure use, several issues will need to be resolved. Extension and training in the use of green manure on a massive scale has not been carried out yet. The production of seed, especially on site, is an area of great importance that has not been worked out, with the exception of *Sesbania* for rice and some of the legumes for forage. Finally, practical on-farm development, incorporating these crops into common farming practices at medium and large-scale production sites will be required.

Zeolite

Zeolite is a naturally occurring mineral, which can be ground and used for many different purposes. Cuba has large zeolite reserves that are mined for agriculture and other sectors. It is rich in calcium, magnesium, potassium, and phosphorous and has a very high cation exchange coefficient (CEC). It not only serves as a natural mineral fertilizer, but also improves nutrient and water retention and availability, activating inert soils and improving soil biochemistry. It produces the best results in highly eroded or degraded soils and can produce yield increases of up to 20 percent.

Zeolite can be used by itself or mixed with organic matter, depending on local soil conditions. Research with natural zeolites in the Mixed Food Crop Enterprise of Manacas municipality in Villa Clara shows that optimum application rates of 6 t/ha of zeolite produce significant physical and chemical improvements in the soil. Huge increases in pH, Ca, Mg, Na, K, CCB, P_2O_5, and K_2O are observed, enhancing yields from 17.4 to 20.5 t/ha in sweet potato crops, 1.9 to 3.4 in garlic, and 8.6 to 21.0 t/ha in tomatoes (Febles 1998).

Minimum Tillage and Animal Traction

As was shown in Table 2, soil compaction from excess mechanical tillage and the use of inappropriate machinery such as disc furrowing is a major problem in Cuba. Additionally, these practices have very high energy costs, and over time reduce organic matter in the soil. In Cuba today there is a dramatic increase in the use of animal traction and of more appropriate machinery, such as the multi-plow, which does not invert the soil layers. Changes in the type and intensity of soil tillage have significant impacts on physical and biological properties of soil. Table 14 offers an example of the effect of mechanization on physical properties of soil.

Table 14. Effect of mechanization on the soils of Pinar del Río province

PROPERTIES	ZERO TILLAGE	TRADITIONAL TILLAGE	MINIMUM TILLAGE	CONVENTIONAL TILLAGE
Apparent density g/cm3	1.2	1.3	1.5	1.6
True density g/cm3	2.4	2.5	2.6	2.7
Total porosity (%)	52	50	45	37
Moisture (%)	18	16	15	10
H4 %	30	30	28	23
% of stable aggregates (>0,25mm)	43	42	40	36
Organic matter	4.9	2.6	2.3	1.9

Note: Traditional tillage—with animal traction; conventional tillage—with tractors

(Vento and Ponce 1998, cited by Febles 1998)

Agroecological Soil Management

Although the term agroecology has only recently entered widespread use in Cuba, agroecological practices are known empirically by many *campesinos.* Many farmers have used or are using crop residues, crop rotation, polycultures, organic manure, crop/livestock integration, and functional biodiversity. Other farmers and some agricultural professionals still subscribe to the agricultural paradigm of the so-called Green Revolution.

Agroecological consciousness in this country has grown considerably in a relatively short number of years, thanks to the massive training and retraining of professional staff and farmers in diverse programs at universities and research centers. The use of the techniques discussed in this chapter must be approached in a more integrated manner, moving from a simple input substitution model towards truly agroecological soil management. This will require a deep-rooted paradigm shift, already underway, allowing agronomists and farmers to view the soil as a living subsystem of an agricultural ecosystem that operates according to the laws of nature.

Encouraging on-farm results have already been obtained, especially in the cooperative sector. One example is the work done to monitor the effects of ecological management of three farms (from 1995 to 1998) in Havana province. This project is part of the SANE-UNDP Agroecological Lighthouse project, with support from the Ministry of Science, Technology, and Environment (CITMA) and the National Association of Small Farmers (ANAP). Over three years soil fertility was maintained and improved in the three farms that were studied (see Table 15).

Table 15. Soil fertility at three Agroecological Lighthouse farms in Havana province

COOPERATIVES (CPA)	YEARS	PH	P, PPM	K, PPM	% ORGANIC MATTER
Gilberto León	1995	7.6	333	183	4.2
	1998	7.7	385	199	4.1
Jorge Dimitrov	1995	7.4	431	292	2.7
	1998	7.7	488	327	3.5
28th of September	1995	6.5	29	265	1.8
	1998	6.7	30	270	3.2

(SANE 1999)

In conclusion, it is clear that soil fertility recovery and improvement is possible in Cuba through ecological management, and that we are well on our way to making this a reality. A considerable amount of relevant research and development is being done, and a tremendous body of knowledge has already been collected. Practical experiences have proven the potential of many of these findings on a commercial scale, and some practices are now commonplace in production. Everyday more people are gaining a new agroecological consciousness about how to make the precious and endangered resource that is the soil produce in perpetuity for the benefit of all Cubans.

References

Arzola, N., V. Paneque, H. Battle, L. Morejón, C. Alfonso, and G. Hernández. 1990. La cachaza como enmienda orgánica y fertilizante para la cañas de azúcar (pamphlet). Havana: INCA.

Bhat, J.B., E.S. Limayen, and B.L. Vasantharajam. 1971. Ecology of the leaf surface microorganisms.

Bustamante, C.R., R. Rivera, M. Ochoa, F. Fernández, M. Rodríguez, and J. Ferran. 1992. Efecto de la aplicación de cepas de hongos micorrizógenos (MVA), bacterias solubilizadoras de fósforo, *Azotobacter chroococum* y niveles de humus de lombriz en el desarrollo de posturas de *Coffea arabica* sobre suelos pardos y ferralíticos. *Resúmenes del I Taller Internacional sobre Biofertilizante en los Trópicos.* Havana.

Cabello, R., L. Rivero, D. Castillo, and J. L. Peña. 1989. *Informe sobre el estudio de la* Sesbania rostrata y S. emerus *como abonos verdes en el mejoramiento y conservación de los suelos arroceros con baja fertilidad.* Instituto de Investigaciones del Arroz. Havana: MINAG.

Cancio, T.M., J.L. Peña, and F.V. Peña. 1989. *Uso de los abonos verdes en áreas tabacaleras de la región del Escambray.* Santa Clara, Cuba: Centro Agrícola, 16 4:59–67.

Cancio, T.M., J.L. Peña, and F.V. Peña. 1990. Aplicación de una labranza reducida y abonos verdes en el cultivo de la cebolla. *Ciencia de la Agricultura* 40:181–183.

CITMA. 1998. *Programa de Acción Nacional de Lucha Contra la Desertificación y la Sequía en la república de Cuba.* Havana: CITMA.

Corbera, J. 1998. Coinoculación *Bradyrhizobium japonicum*–micorriza vesículo–arbuscular como fuente alternativa de fertilización para el cultivo de la soya. *Cultivos Tropicales* 19(1):17–20.

Crespo, G. and O. Gutiérrez. 1992. Estudio de métodos para aumentar la efectividad del estiércol vacuno como abono orgánico. *Resúmenes del I Taller Internacional sobre Biofertilizantes en los Trópicos.* Havana.

Cueto, C.R. and M. Sánchez. 1994. Interacción de cepas de hongos MVA en plántulas de maracuyá (*Pasiflora edulis* Sims var. *flavicarpa*). XVII Reunión Latinoamericana de Rhizobiología. Havana, pg. 103.

Dibut, B. 1998. Efecto de Azotobacter chroococcum sobre el cultivo de la cebolla (dissertation). Havana: UNAH.

Dibut, B.; A. Rodríguez y R. Martínez Viera. 1996. *Efecto de la doble función de* Azotoryza *sobre el plátano* Musa *sp.* INFOMUSA 5 (1):20–23.

Febles, J.M. 1998. Aportes para el manejo ecológico de los suelos en Cuba (pamphlet). Havana: UNAH.

Fernández, F., R. Ortiz, M.A. Martínez., A. Costales, and D. Llonín. 1997. The effect of commercial arbuscular mycorrhizal fungi (AMG) inoculants on rice *Oryza sativa* in different types of soils. *Cultivos Tropicales* 18(1):5–9.

Gandarilla, J. E., D. Pérez, and R.D. Curbelo. 1995. Sistemas bioorgánicos de nutrición para organopónicos. II Encuentro Nacional de Agricultura Orgánica. *Programa y Resúmenes.* Havana.

García, M. 1998. Contribución al estudio y utilización de los abonos verdes en cultivos económicos desarrollados sobre un suelo ferralítico rojo de La Habana (dissertation). Havana: UNAH.

García, M. and E. Treto. 1997. Contribución al estudio y utilización de los abonos verdes en cultivos y utilización de los abonos verdes en cultivos económicos desarrollados sobre suelos ferralíticos rojos en las condiciones de Cuba. *Resúmenes del I Taller*

Nacional de Producción Agroecológica de Cultivos Alimenticios en Condiciones Tropicales. Havana: Liliana Dimitrova Horticultural Research Institute, pg. 74.

Gárciga, M. A. Fernández, and E. Pouyú. 1992. Perspectivas del uso de micorriza en el cultivo del tabaco. *Resúmenes del I Taller Internacional sobre Biofertilizantes en los Trópicos.* Havana.

González, P.J, G. Navarro, D. Fernández, and F. Camina. 1996. La lombricultura: Una opción productiva. *Agricultura Orgánica* 2(1):15–17.

Hernández, G., V. Toscano, H. Vázquez, L.A. Gómez, N. Méndez, M. Sánchez, and M. Mosquera. 1994. Uso y manejo de inoculantes a base de *Rhizobium* en vignas. Cultivos Tropicales. *Programa y Resúmenes.* IX Seminario Científico del Instituto Nacional de Ciencias Agrícolas. Havana.

Hernández, G., J.J. Drevon, H. García, and B. Faure. 1998. Fijación simbiótica del nitrógeno; resultados obtenidos en Cuba para frijol común. *Programa y Resúmenes.* XI Seminario Científico del Instituto Nacional de Ciencias Agrícolas. Havana.

Herrera, R.A. 1994. Ecología de las micorrizas en ecosistemas tropicales. XVII Reunión Latinoamericana de Rhizobiología. Havana, pg. 38.

INCA. 1997. Ecomic, Azofert y Rhizofert, biofertilizantes originales, seguros y eficientes para la agricultura. Havana.

López, M. 1994. La biofertilización en las leguminosas de pastos y soya en Cuba. XVII Reunión Latinoamericana de Rhizobiología. Havana, pg. 36.

Martínez Viera, R. 1997. Los biofertilizantes como pilares básicos de la agricultura sostenible en Cuba. *Conferencias del I Taller Nacional de Producción Agroecológica d Cultivos Alimenticios en Condiciones Tropicales.* Havana: Liliana Dimitrova Horticultural Research Institute, pg. 88.

Martínez Viera, R. and B. Dibut. 1996a. Beneficios de la utilización de los biofertilizantes en Cuba. *Memorias del I Encuentro Internacional sobre Agricultura Urbana y su impacto en la comunidad.* Havana, pp. 61–67.

Martínez Viera, R. and B. Dibut. 1996b. Los biofertilizantes como pilares básicos de la agricultura sostenible. *Curso–Taller Gestión medioambiental de Desarrollo Rural.* Havana, pp. 62–81.

Martínez Viera, R. and G. Hernández. 1995. Los biofertilizantes en la agricultura cubana. *Resúmenes del III Encuentro Nacional de Agricultura Orgánica, Conferencias.* Havana, pg. 43.

Mayea, S. 1994. *Tecnología para la producción de compost (biotierra) a partir de la inoculación con microorganismos de diversos restos vegetales.* Havana: CIDEA.

MINAG. 1993. *Instructivo técnico de lombricultura.* Havana: Delegación Provincial del MINAG.

Nardo, A., E. G. Cañizares, M. García, J. Azcuy, J. Sosa, J.M. Calaña, and M. Guerra. 1992. Evaluación de la influencia de dos tipos de biofertilizantes en el crecimiento y desarrollo de posturas de mandarino cleopatra en la fase de vivero. *Resúmenes del I Taller Internacional sobre Biofertilizantes en los Trópicos.* Havana.

Noval, B. de la, M. Hernández, and J.C. Hernández. 1997. Utilización de las micorrizas arbusculares en la adaptación de vitroplantas de banano (*Musa* sp.) dosis y cepas de hongos formadores de MVA y combinaciones de sustratos. *Cultivos Tropicales* 18(3):5–9.

Noval, B. de la, F. Fernández, and R. Herrera. 1995. Efecto del uso de micorriza arbuscular y combinaciones de sustrato sobre el crecimiento y desarrollo de vitroplantas de piña. *Cultivos Tropicales* 16(1):19–22.

Paneque, V.M., P.J. González, and J.M. Calaña. 1986. *Estudio de los residuos del Centro de Acopio "La Paulina" del Complejo Agroindustrial Azucarero "Abraham Lincoln."* Havana: Servicio Científico-Técnico del Departmento de Agroquímica del INCA.

Paneque, V.M., P.J. González, and J. Fernández. 1991. Utilización de las aguas residuales del CAI "Juan Manuel Márquez" para el riego y la fertilización de la caña de azúcar. *Cultivos Tropicales* 12(2):5–8.

Paneque, V.M. and M.A. Martínez. 1992. *Evaluación de la vinasa del CAI "Héctor Molina" como fertilizante para la caña de azúcar.* Havana.

Paneque, V.M. and G. Grass. 1992. Caracterización de las aguas residuales del combinado de Cítricos Jagüey Grande y su potencialidad como fuente de riego y fertilización (pamphlet). Havana: INCA San José de las Lajas.

Paneque, V.M. 1999. Utilización de los residuales de la industria azucarera en el fertirriego de la caña de azúcar (pamphlet). MINAZ.

Peña, J.L., F. Peña, and T. Cancio. 1986. Comportamiento del millo *Sorghum vulgare,* Pers. Como cultivo antierosivo y mejorador del suelo en áreas de producción de semillas de papa en la montaña. *Jornada Científica de la Est.* Escambray, Barajagua, Cienfuegos.

Peña, J.L., F. Peña, T. Cancio, and B. Gronzo. 1988. Instructivo técnico sobre el uso de los abonos verdes como protectores y mejoradores del suelo. Havana: MINAG.

Peña, E., M. Carrión, and R. González. 1995. La cachaza como sustrato en organopónicos. *II Encuentro Nacional de Agricultura Orgánica: Programa y Resúmenes.* Havana.

Pijeira, L. and E. Treto. 1983. Estudio del comportamiento de las cepas de *Rhizobium japonicum* asociadas a variedades de soya de primavera. *Cultivos Tropicales* 5(1):61–73.

Pijeira, L., E. Treto, J.D. Mederos, J. Corbera, A. Velazco, M. Castellanos, and N. Medina. 1988. La nutrición y fertilización de la soya cultivada en condiciones de un suelo ferralítico rojo compactado de Cuba. *Cultivos Tropicales* 10(3):19–26.

Ramón-Cuevas, J., O. Morejón, M. Ojeda, and V. Vale. 1996. La lombricultura. 1. Una opción ecológica. *Agricultura Orgánica* 2(1):13–14.

Rivera, R., F. Fernández, C. Sánchez, C. Bustamante, R. Herrera, and M. Ochoa. 1997. Efecto de la inoculación con hongos micorrizógenos V.A. y bacterias rizosféricas sobre el crecimiento de las posturas de cafeto. *Cultivos Tropicales* 18(3):15–23.

Roldós, J., F. González, M. Pérez, E. García, J. Hernández, A. Gil, and A. Menéndez. 1992. La aplicación de biopreparados a base de *Azospirillum* y su efecto sobre la productividad de la caña de azúcar. *Resúmenes del I Taller Internacional sobre Biofertilizantes en los Trópicos.* Havana.

Ruiz, L. 1993. Factores que condicionan la eficiencia de las micorrizas arbusculares, como alternativa para la fertilización de las raíces y tubérculos tropicales (dissertation). Havana: UNAH.

SANE. 1999. *Proyecto del PNUD* (final report). Havana: Grupo Gestor de Agricultura Orgánica.

Treto, E., M. García, R. Brunet, J. Herrera, J. Kessel, R. Gómez, R. Iglesias, and H. Santana. 1992. Nutrición y fertilización de la piña, 20 años de investigación en el Instituto Nacional de Ciencias Agrícolas. *Cultivos Tropicales* 13(2–3): 5–59.

Treto, E., M. García, M. Alvarez, and L. Fernández. 1994. Abonos verdes: algunas posibilidades de su uso en la agricultura cubana. IV Forum Nacional de Ciencia y Técnica. Havana.

Velazco, A., R. Castro, M.C. Nápoles, F. Cuevas, G. Díaz, and T. Hernández. 1992. Uso del *Azospirillum brasilense* en el cultivo del arroz. *Resúmenes del I Taller Internacional sobre Biofertilizantes en los Trópicos.*

CHAPTER ELEVEN

The Integration of Crops and Livestock

Marta Monzote, Pasture and Forage Research Institute, Eulogio Muñoz, Animal Science Institute, and Fernando Funes-Monzote, Pasture and Forage Research Institute

As with other agricultural sectors in Cuba, cattle raising was highly specialized as part of the intensive production model. Production levels were maintained through the importation of large quantities of inputs. For many years the Council for Mutual Economic Assistance (COMECON, economic union of the former Socialist Bloc) facilitated the acquisition of these inputs through a trade regime that benefited Cuban agriculture. Livestock production was developed with an eye toward guaranteeing milk consumption, among other products, as a basic diet for children, the elderly, and the sick, as well as being a component of the diet for the rest of the population. Nevertheless, the separation of crop and livestock production that took place was wasteful of energy and nutrients.

The "Special Period" triggered a search for alternative means of livestock production using local resources. Luckily, despite the specialization of the earlier period, many small and medium-sized farmers had preserved and enhanced traditional mixed systems of land use, based on locally available resources, which might serve as a starting point.

Positive research results were rapidly obtained by substituting biological inputs for chemical inputs, and by using local instead of imported feed sources. This was a step toward a scientific understanding of integrated crop–livestock production systems.

Industrial Livestock Production

Cuba's livestock production before the 1959 Revolution was extensive in nature, with little infrastructure, and was mainly focused on beef production. With the Revolution this changed to intensive-style production with heavy use of inputs and an infrastructure geared toward milk production.

A new breeding program was based on the transformation of Cebu cattle (a local breed) using Canadian Holstein bloodlines. The cattle resulting from this cross have adequate production levels under tropical conditions, good adaptation to Cuba's climate, and lower requirements for high-quality feeds than pure Holstein cattle. They can be subjected to low-input systems where they are still able to express their productive potential, and are also used as double-purpose cattle for dairy and beef production. Artificial insemination (AI) was widely employed in this program, with eight AI centers and more than 900 sires and some 2,500 inseminators (Anon 1987).

In the 1960s the state determined the principle source of feed for cattle should be improved pastures. Thus a huge program was initiated of planting improved plant varieties, and new species were introduced (see Table 1) over an area that covered almost one million hectares. However, the productive life of these improved pastures was not more than five years, mostly due to poor management.

Table 1. Principle species used for pasture improvement

GRASSES	LEGUMES
Pangola grass (*Digitaria decumbens*)	Perennial soybean (*Neonotonia wightii*)
Guinea grass (*Panicum maximum*)	Leucaena (*Leucaena leucocephala*)
Bermuda grass (*Cynodon dactylon*)	Stylo (*Stylosanthes guianensis*)
Star grass (*Cynodon nlemfuensis*)	Tropical kudzu (*Pueraria phaseoloides*)
King grass (*Pennisetum purpureum*)	Alfalfa (*Medicago sativa*)
Signal grass (*Brachiaria* spp.)	
Rhodes grass (*Chloris gayana*)	

Cuban researchers had shown it possible to obtain 8 to 12 liters of milk per cow on a diet of just Cuban pasture (Jeréz et al. 1986, García Trujillo 1993, and Funes and Monzote 1993), but the livestock industry used large amounts of imported concentrated feeds (600,000 tons), mineral salts (25,000 tons), fish meal (36,000 tons), urea, and other supplements each year. During the 1980s there were 3,100 modern dairy units with mechanical milking and advanced animal management methods. Infrastructure included roads, electrification, irrigation networks, and 17,300 tractors and allowed for the expansion of intensive cattle production into the most remote areas. Modern technology simplified work in the countryside, and maximum milk production of more than

Ubre Blanca (White Udder), world-record milk producer, with her calf.

900 million liters of milk per year was obtained. Although cattle production was not efficient, it did represent 14 percent of GDP (Anon 1987).

At this stage, average yield did not exceed 6.3 liters of milk per cow, which is below the genetic potential of the animals (although isolated herds yielded more than 20 liters of milk per cow per day). The world record was held by the Cuban cow *Ubre Blanca* (White Udder), who yielded 110.9 liters of milk in one day and 27,000 liters over a total lactation.

Dairy and beef production levels were attained at a high energetic cost. Funes-Monzote (1998) analyzed energy efficiency during the peak of milk production, and found that 5.7 calories were needed to produce one calorie of milk (see Table 2).

Energy efficiency in Cuba is closely linked to economic efficiency. This is due mainly to the scarcity of fossil fuels, technologies, and capital for high-input cattle production. The Fifth Congress of the Cuban Communist Party in 1998 (Anon 1998) recognized this, in stating that, "as part of economic efficiency, emphasis should be placed on labor discipline and technology, as well as on energy and productive efficiency, among others. It would be necessary not only to reach these goals, but to quantify them in order to know how much could have been done and compare this to what was achieved."

During the era of the industrial agricultural model, one of the most distinctive characteristics was that it was highly dependent on imports, and did not fully meet human and animal feed requirements. Under these conditions cattle production was unsustainable.

Table 2. Estimation of livestock production energy efficiency in Cuba at peak production levels

ENERGY PRODUCTION			
PRODUCT	PRODUCTION, kg	ENERGY EQUIVALENTS (kcal/kg)	ENERGY PRODUCTION (millions Mcal)
Milk	900 million	660	594
Beef	300 million	2,280	684
Subtotal (production)			1,278
ENERGY COSTS			
Fuels	115,000 tons	9,243	1,063
Nitrogen fertilizers:			
Nitrogen (N)	229,700 t	12,300	2,825.3
Formulas:			
Nitrogen (N)	12,357 t	12,300	152
Phosphorus (P)	7,414 t	1,000	7.4
Potassium (K)	14,828 t	1,200	17.8
Feed concentrates	600,000 t	2,460	1,476
Molasses	870,000 t	2,120	1,844.4
Labor	100,000 men	250	63.9
Subtotal (costs)			7,516.4
Energy balance = 0.17 calorie produced/calorie input			

Toward Food Self-Sufficiency

The 1970s and 1980s

In the late 1970s, researchers were faced with the challenge of making milk and beef production more efficient. Cuban research centers refocused their efforts on the following options: grass/legume relationships, protein banks, production and utilization of biological pesticides, biological fertilizers, adaptation and evaluation of low-input pasture varieties, minimum tillage for grass and legume establishment, cattle manure as fertilizer, silvipastoral methods, and nutritional blocks (Monzote and Funes-Monzote 1997). In the 1980s, preliminary results were obtained from these research programs (see Table 3).

New cattle feeds from crop by-products, harvest residues, and nonconventional ingredients were used to substitute for some imported inputs for livestock production (see Table 4). Sugarcane was a major contributor, with more than four million tons a year used in eleven different types of feeds, more than two million tons of which were consumed by cattle.

Pasturing cattle in association of grass and legumes.

Table 3. Scientific results on pasture improvement during the 1980s

TOPIC	RESULTS	AUTHORS
Manure utilization for pasture improvement	Evaluation of bovine, pig, and poultry manures as fertilizer for pastures. Keeping cattle manure exposed to fresh air increased assimilation by plants, after three months, from 46.6% to 63.7%.	Crespo and Hernández 1985, Crespo 1984, Arteaga et al. 1986
Grass/legume pasture mixtures	Establishing associations by planting legumes in grasslands using minimal tillage. The protein content of grasslands increased by 5%, yield by 30%, digestibility reached 66%, mineral and vitamin content increased, and soil quality improved. Beef production of from 480 to 650 g/animal/day, and milk yields increased from between 1 to 2 liters/cow/day.	Monzote 1982
Legume protein banks	Increases of one to two liters of milk/cow/day and better stability of the legumes.	Funes and Jordán 1987, Marrero 1989
Silvipastoral methods	7 to 13 sheep/ha in citrus orchards produced 343 to 592 kg/LW/ha/year without reducing yield and quality of citrus fruit. LW = live weight	Borroto 1988
Forage for self-provisioning of feed	Sugarcane works as energy supplement in feed, but a protein source must also be included in the diet. Certain varieties of sugarcane were selected as forage.	Stuart et al. 1985
Biological control	*Bacillus thuringiensis, Metarhizium anisopliae,* and *Beauveria bassiana* for the control of *Mocis latipes* and *Monecphora bicinta fraterna* on improved pastures. *B. bassiana* and *M. anisopliae* strains for the control of cattle ticks (*Boophilus microplus*).	Martínez-Mojena 1986, Castiñeiras et al. 1986
Biofertilizers	*Rhizobium* strains increased pasture production by 30% and fixed 250 kg/ha of nitrogen. *Leucaena* outperformed chemical fertilization by 4 tons DM/ha. DM = dry matter	López 1981, Tang 1986
Selection of pasture species	Species and varieties of introduced and native grasses and legumes adapted to different regions.	Funes et al. 1971a, 1971b, 1986; Menéndez 1982; Hernández 1989

By-products from the sugar industry have been used for feeding animals in Cuba since the early 1960s (Pérez-Infante and García-Vila 1975). Among these by-products are molasses, bagasse, *cachaza* and, in the 1980s, *saccharina* (obtained from the fermentation of ground sugar cane with urea) (García et al. 1994). For many years animals have also been transported to sugarcane collection centers to consume fresh cane residues. However, many research findings had less widespread adoption.

Despite promising results from research and small-scale production units, there were several reasons for the low level of commercial-scale application of the majority of these results. The inflexible structure and organization of the conventional model of agriculture, and the advantageous trade relations with the Socialist Bloc—making imported feeds relatively cheap—made the implementation these advances difficult, so that in many cases research results remained at the experimental or semicommercial level.

Table 4. Research and development on animal feed production in the 1980s

OVERALL TOPIC	METHODS	FEED SOURCES
• Evaluation and development of crop residues, by-products, and other nonconventional ingredients for animal feed	• Biological enrichment • Processing and enrichment using physical and chemical methods • Rational diet formulation for different species and domestic animals • Industrialization of feed production	• Sugarcane: tops and crop residues, bagasse, molasses, *cachaza,* and other by-products • Crop residues and milling residues from domestic cereals like rice • By-products of citrus and other fruit processing • Other resources from agriculture, including coffee residues, cocoa, coconut, kenaf, roots and tubers, etc. • Manure and slaughter-house wastes

The 1990s

Livestock production suffered severe impacts from the collapse of the Socialist Bloc. The availability of imported and nationally produced feed resources dropped drastically. The reduced level of imports (see Table 5) made the conventional high-input model (which had in the past made it possible to reach acceptable production levels) collapse, showing the vulnerability of this production system. For instance, in Havana province where dairy operations were highly specialized compared to the rest of the country, milk production fell from 320 to 60 million liters of milk per year. This led to a sharp deficit in milk supply to the internal market.

In light of the new situation, those responsible for the technological, and socioeconomic aspects of cattle production promoted the implementation of certain measures to help reduce losses (Perón and Márquez 1992). Among them were:

- a priority on pasture utilization (rather than concentrated feeds)
- on-farm rearing of replacement cattle for those lost initially
- on-farm production of animal feed and fodder
- replace artificial feeding of calves with suckling from mother cows
- redirect cattle breeding toward locally adapted breeds
- build housing facilities in the production units
- produce food for farm families on the farm
- decentralization of large enterprises

Table 5. Drop in availability of inputs for animal feed and pasture improvement

ITEM	MAXIMUM IMPORT LEVEL (1000s METRIC TONS)	1995	% REDUCTION OVER PRE-1990 LEVELS
Feed concentrate	563.7	37.8	93
Protein supplements	68	8.2	88
Urea	90	3	98
Mineral salts	25	0.7	97
Molasses	1,090	226.6	79
Nitrogen fertilizers	229.9	–	100
NPK fertilizers	211	–	100
Silage	4,123 (green matter)	–	100
Hay	1,660 (green matter)	250	85

(Statistical bulletin, MINAG)

Many years earlier Ugarte (1972) had shown that both pure Holsteins and the Holstein x Cebu cross produced more milk, and their offspring gained more weight, if they were able to nurse them themselves, compared to artificial feeding of calves. The scale of operation proved to be important under the new conditions, as mostly Holstein cows in Havana province mini-dairies in 1994 did much better than those in large production units (see Table 6).

Table 6. Comparative performance of small and typical dairy units, Havana province, 1994

INDICATORS	MINI-DAIRIES	TYPICAL DAIRY UNITS
No. of production units	595	910
Cows/unit	24	66
Workers/unit	1.7	6.5
Milk production		
Liters per lactating cow per day	7.2	2.7
Liters/cow/year	1878	610
Liters per cow (total) per day	4.1	1.3
Birthrate, %	60.0	50.2
Calf mortality, %	6.0	20.8
Total mortality, %	4.6	15.5

(Martín et al. 1996)

The creation of the Basic Units of Cooperative Production (UBPC) brought many changes to land tenure and cattle ownership, and led to reductions in the average size of cattle management units. Already by 1995, 730 UBPCs were engaged in cattle production, accounting for 26.5 percent of the national cattle herd, and contributing 45 percent of total milk production (see Table 7).

Table 7. Distribution of cattle by property type in 1995

TYPE OF PROPERTY	% OF TOTAL CATTLE
State farms	47.5
UBPCs	26.5
CPAs	7.0
Individuals	10.0
Others	9.0
Total	100.0

(ONE 1995)

Law 142, which set up the UBPCs (MINAG 1997), reduced the size of cattle holdings and provided incentives for product diversification. Article 1 of the law states:

> *b) The labor collective should produce basic foodstuffs in addition to products for market, in order to raise the level of self-sufficiency of the farm.*
>
> *c) Resources should be managed to maximize the self-sufficiency of the production unit.*

Similarly, Article 60 of the fourth section of resolution No. 354/93 (which set rules for UBPCs) states that each unit "should work towards the diversification of its production while not affecting its main product line." Chapter five touches upon reforestation, declaring that "the enterprise and the UBPC should prepare a program for the creation of forest and fruit plantations to satisfy requirements of the UBPC and the community."

Within the larger policy framework designed to recover earlier levels of cattle production, high priority is now given to the production of all necessary feeds within the boundaries of the production unit. Typically this includes areas devoted to sugarcane, legumes (mainly *Leucaena*), king grass, and other forages and fodders, as sources of animal nutrition.

Even though these changes have been made, there is still a further need for agricultural and livestock diversification. The full productive potential of crop/livestock integration has yet to be developed on any significant scale. There is still a long way to go toward taking full advantage of manure as fertilizer for crops and the non-harvest biomass of the crops as feed for animals, with the exception of processing residues from sugarcane, citrus, and rice, and these are sometimes transported over long distances.

Over the next few years the plan is to use 121,000 tons of wet and dry citrus residues for animal feed in Havana, Havana province, the Isle of Youth, Matanzas, and Ciego de Ávila (MINAG 1998). Rotational systems for fattening cattle with rice stalks after harvest have already been widely used in Pinar del Río, Sancti Spíritus, and Granma provinces with satisfactory results.

By 1993, the new agricultural policy had prevented further reductions in milk production as recovery began (see Figure 1 and Table 8), and losses to the cattle population were ended, thanks in large part to the distribution of agricultural land and animals to the private sector (Rodríguez 1998), where farms were smaller and more diversified.

Figure 1. National milk production 1991–1999

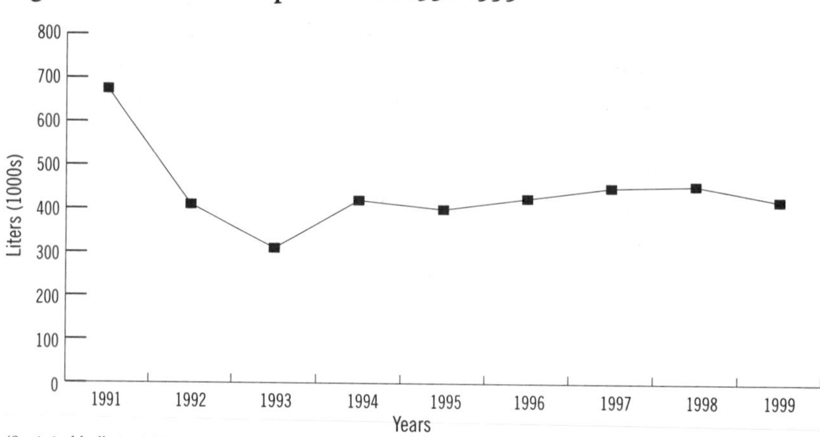

(Statistical bulletin, MINAG)

Table 8. National milk production (millions of liters), 1996–1999

SOURCE OF PRODUCTION	1996	1997	1998	1999*
State farm	115.7	117.6	110.9	251.7
UBPC	193.3	190.8	180.7	–
CPA	49.2	47.8	48.2	168.0
CCS	68.1	89.6	114.0	–
Total	426.3	445.8	453.8	419.7

*Sum state farms + UBPC + CPA + CCS

(Statistical bulletins, MINAG)

It should be noted that with the new self-provisioning areas for the workers at many kinds of businesses and institutions, the increase in the number of individuals with small farms, and the production coming from the Young Workers' Army (EJT),—all largely unreported—actual levels of production in contemporary Cuba are more substantial than official figures show. However, there is no doubt that per animal production and per unit area production is still relatively low. This does not coincide with scientific results achieved in low-input milk production, demonstrating that considerable additional production increases are possible under present conditions.

Crop/Livestock Integration and Agroecology

Agricultural and livestock integration is a slow process due to the high degree of specialization Cuba achieved under the conventional model. Although the process of integration has been initiated, it is still being carried out mostly with an eye toward meeting self-provisioning needs, and frequently follows the old intensive agriculture pattern (García Trujillo and Monzote 1995). Organic agriculture techniques and methods provide for sustainable feed production through more integrated systems of livestock and crop production (Muñoz et al. 1993). Such integration, if carried out over a large scale (i.e. trucking manure over long distances, etc.), would involve heavy expenses in terms of resources, mainly fuel, which would negatively affect the economic and energetic efficiency of production, especially in countries like Cuba with scarce energy sources. For this and other reasons, the sustainable agriculture research community, inside and out of Cuba, focuses on the potential of systems based on integrating the animal/crop relationship within the farm.

Many studies on this topic agree that via maximized nutrient recycling, conservation, and/or recovery of on-farm natural resources, and the establishment of general order in the farming system, efficiency, stability, and increased levels of production can be attained. However, this has often been a hard argument to make in Cuba, as good data showing this was previously lacking (Funes-Monzote 1998).

Horses pastured in citrus orchards, "Victoria de Girón" Citrus Enterprise, Matanzas province.

More recently, researchers at the Pasture and Forage Research Institute (IIPF), the Animal Science Institute (ICA), the "Indio Hatuey" Pasture and Forage Experiment Station (EEPF), and the agricultural universities and other institutions have carried out a series of experiments to study more fully integrated production systems. This work has been financed by the Ministry of Science, Technology, and Environment (CITMA), and by the Ministry of Agriculture (MINAG).

In one such effort the "Indio Hatuey" Station in Matanzas province conducted research on integrated systems involving grazing horses in orange plantations. The experiments have lasted for several years with good results (Simón and Esperance 1997). The horses did not damage the orange trees, and saved labor and/or herbicide normally used to clear weeds from the orchard floor, a savings of 219 Cuban pesos/ha/year in salaries, fuel and/or herbicide. Horse grazing reduced guinea grass (*P. maximum*) and *faragua* (*Hyparrhenia rufa*) populations in a magnitude related to the density of horses. This favored the growth of grasses and legumes that are less competitive with the orange trees. The animals also recycled 2 t/ha/year of organic matter, and furnished 40, 42, 12, and 51 kg/ha of N, P, K, and Ca, respectively. From an economic viewpoint, the orange/horse integrated system yielded a profit that was 388 Cuban pesos/ha/year higher than the orange monoculture without animals. This experiment station has also carried out many trials on the introduction of silvopastoral systems based on *Leucaena leucocephala* in livestock enterprises in Havana and Matanzas provinces.

Since 1994 the Pasture and Forage Research Institute has carried out research on integrated crop/livestock systems based on agroecological principles. An example is the "System Design for Crop/Livestock Integration on

Small and Medium-Sized Farms" project, which was carried out in seven provinces, as part of the "National Program of Food Production through Biotechnological and Sustainable Means" of CITMA. The initial project has concluded, but research continues through an agroecological network created by the pasture research stations, which bring together *campesinos* and other farmers who use the principals of crop/livestock integration on their farms (Monzote et al. 1999).

Because the higher productivity of integrated systems is the result of complex mechanisms, it is common in this type of research that ones obtains extremely positive results without being able to provide a clear explanation of their causes. In general terms, it is well known that crop diversification and nutrient recycling in integrated crop/livestock systems create synergisms which facilitate the productive capacity of each component of the system. Among other advantages often found in complex systems are a reduction of vulnerability to pests, diseases, and weeds, a lower dependency on external inputs, lower capital requirements, and a greater efficiency of land use (Rosset 1998).

People around the world are studying how to analyze complex agroecological systems more effectively; certain methodologies already allow for the interpretation of results. Various indicators have been used to evaluate production efficiency, though these do not reflect the sustainability aspects of such systems. Socioeconomic data, together with production and efficiency indicators, and measures of soil conservation and functional biodiversity should be considered, with policies tailored to local conditions. In evaluating crop/livestock systems in Cuba, we must also take into consideration deforestation/reforestation, utilization of manure for soil fertilization, fuel efficiency, and labor requirements (as labor is scarce in rural areas) (Funes-Monzote and Monzote 2000).

Energy and Production Efficiency

Energy efficiency is an indicator of productivity in agricultural systems and is expressed as the amount of calories produced per calorie consumed. This term and its use to evaluate the efficiency of systems have been problematic. Nevertheless, it is an indicator of sustainability which allows for comparisons, and establishes levels of efficiency as a function of the two main variables in a productive system, inputs and outputs (Funes-Monzote et al. 2000).

On implementing these integrated crop/livestock systems on a pilot basis, the productive results surpassed those of the specialized livestock system that previously prevailed in Cuba. The pilot systems reached equal or higher milk yields on smaller areas dedicated to cattle raising. Also, many agricultural products for human consumption were produced, all of which make the livestock system more profitable and create an important surplus of sellable resources.

This is important due to the scarcity and high prices of these products in the market.

The milk production obtained in these systems—yielding one to three tons per hectare of the *entire* farm, without counting that consumed by calves—implies that production on the portion of the farm actually devoted to grazing is up to six tons per hectare, which is very high for tropical systems. This was achieved with very few external inputs, as shown in Table 9 (Funes-Monzote et al. 2000).

Table 9. Performance and energy efficiency in integrated crop–livestock pilot farms over three years

INDICATOR	RANGE
Observed	
Area (hectares)	1–20
Total production, all products, tons/ha	4–9
• Crop production (t/ha)	3–6
• Livestock production (t/ha)	1–3
Energy (Mcal/ha)	3,000–10,000
Protein (kg/ha)	100–300
Individuals fed/ha	
• Energy sources	4–9
• Protein sources	
1. Plant origin	3–10
2. Animal origin	5–12
Inputs (energy expenses)	
Labor force (Mcal)	500–1,000
Animal force (Mcal)	20–60
Tractors	0–300
Energy ratio (calories produced/calories input)	2–10

(Monzote et al. 1999)

On examining the performance of the integrated farms, a constant increase over time of production and of energy efficiency was evident. Table 10 provides an evaluation of one farm over time, based on productivity/area and the contribution of protein and energy as components of the human diet (Sosa and Funes-Monzote 1998).

Table 10. Evaluation of a 75:25 (livestock area/crop area) integrated farm

INDICATOR	FIRST YEAR	SECOND YEAR	THIRD YEAR	FOURTH YEAR
Area (ha)	1	1	1	1
Total production t/ha	4.4	4.9	5.1	5.3
• Plant production (t/ha)	1.3	3.3	2.8	2.4
• Livestock production (t/ha)	3.1	1.6	2.3	2.9
Energy (Mcal/ha)	3,797	3,611	4,885	4,393
Protein (kg/ha)	168	115	151	147
Individuals fed/ha				
• Energy sources	4.0	3.5	4.8	4.3
• Protein sources				
1. Plant origin	3.0	2.4	2.2	2.4
2. Animal origin	12.0	8.0	11.4	10.7
Inputs (energy expenses)				
Labor force (Mcal)	569	392	359	316
Animal force (Mcal)	16.8	16.8	16.8	1
Tractors	277.3	–	138.6	46.2
Energy ratio (calories produced/calories input)	4.4	8.8	9.5	12.1

(Funes-Monzote and Monzote 2000)

Biodiversity

If the starting point of a process of bio-diversification is a specialized dairy farm, then results can be noted beginning in the first year. The simple insertion of some crop production into these systems increases biodiversity—expressed mainly in the number of cultivated plants, vegetables, medicinal plants, herbs, and fruit trees present, all new components of the functional biodiversity of the farm. The increased production of foodstuffs is also an indicator of biodiversity, often ranging from 18 to 27 products once a production unit begins the integration process.

The greater number of species of natural enemies of pests found on more integrated farms, indicates a multiplier effect of cultivated plant diversity on overall biodiversity and on the ecological control of pests and diseases. This is part of the enhanced bio-regulatory capacity and stability of these systems (Pérez-Olaya 1998).

The higher values of biological activity in the soil biota reveal an intense decomposition of plant material by beneficial organisms, especially earthworms, mealy bugs, and other fauna. The biological activity of compost produced, measured by net respiration, is almost three times greater than that of the soil near the surface (0–10 cm), and four times greater than soil at 10–20 cm (Rodríguez 1998).

Reforestation is also part of functional biodiversity on integrated farms. This includes reforestation in both the grazing and crop areas (Monzote and Funes-Monzote 1997). In the grazing areas this takes the form of tree planting in pastures as well as living fence posts. Crop areas involve tree planting in strips between crop rows—like alley cropping—as well as along perimeters.

Reforestation is no easy task, because the animals initially will damage seedlings; this effect has been reduced by planting nonpalatable wood and fruit trees in livestock areas, and/or by protecting seedlings.

Organic Fertilizers

An important aspect of these systems is the utilization and transformation of manure into organic fertilizers for crops. The objective is to avoid contaminating the environment with runoff, while improving productivity, sustainability, and nutrient recycling (Funes-Monzote and Hernández 1996). Part of the nutrient recycling takes place via compost made from residues of maize, beans, cassava, plantains, sweet potatoes, grass, and *Leucaena*. Manure and the rejected materials from animal feeding (sugarcane, ground *Leucaena*) represent additional sources of organic matter for composting.

In Cuba, livestock systems have generally been developed under relatively dry conditions. This lends itself to the use of sugarcane as forage during the dry season in most production units (cassava is also used). The cane rejected by the animals and cassava by-products have been used for compost with very good results (see Table 11).

Table 11. Quality of compost prepared with different raw materials

INGREDIENTS	HUMIDITY %	OM %	N%	P%	K%	PH
Manure Ground cassava stems Ground sugarcane	20.8	65.5	2.7	1.3	1.9	6.3
Manure Ground sugarcane	15.9	52.2	1.0	0.7	1.8	6.4
Ground sugarcane Ground cassava stems	13.5	44.4	1.1	0.6	0.8	7.3

(Monzote et al. 1999)

In organic farming systems based on agroecological principles it is not considered ecologically acceptable to make heavy use of off-farm, external inputs. Instead, alternatives such as green manure, mulch, compost, and worm humus are used as substitutes for chemical fertilizers and can be obtained or produced on-farm (see Table 12). The particular balance between different sources of organic fertilizer depends on the design of the farm. Integrated production systems with higher proportions of area devoted to crops produce more biomass to incorporate. Where animals dominate, more manure is available.

Table 12. Organic fertilizers produced per hectare on a 50–50 crop–livestock integrated farm

TYPE	FIRST YEAR (1996)	SECOND YEAR (1997)	THIRD YEAR (1998)
Compost, (t/ha)	0.2	1.8	6
Area improved (m^2)	250	600	2,000
Mulch (t/ha)	6.8	8.7	8.6
Area improved (m^2)	3,800	3,190	2,600
Green manure (tDM/ha)	5.6	4.5	4.2
Area improved (m^2)	2,380	2,200	2,600

(Monzote et al. 1999)

The utilization of compost and other organic fertilizers in specialized monoculture systems, and/or on large-scale production units, has a series of drawbacks. Limiting factors include the availability of raw materials, high transport and application costs, and the labor and equipment required. However, when the scale of the system is kept smaller, and the degree of integration high, using these techniques is much easier, and in fact becomes a functional necessity of the system, while guaranteeing nutrient recycling.

Human food waste, crop residues, animal feeding leftovers, and excess hay can all be profitably used and transformed into enriched organic fertilizers for crops by stimulating natural decomposition via microorganisms and soil biota. This should be practiced at every level and scale of production.

Labor intensity

The labor intensive nature of organic farming on a small or medium scale is one of its most problematic aspects. However, we found that on one hectare integrated farms with up to 50 percent of the area in crops, the average labor intensity can be as low as 4 hr/day at the start, gradually decreasing with the establishment of the farm. This indicator should be carefully studied, since it is related to the socioeconomic conditions of each region, the intensity of production, and the scale and the level of mechanization employed.

Evaluating the sustainability of integrated farms

The joint interpretation of indicators allows us to evaluate the sustainability of agroecological systems. Radar or "kite" graphs visually summarize indicators from the farms and make it easier to evaluate sustainability patterns (see Figure 2).

Figure 2. Sustainability indicators measured on integrated farms with different proportions of crops and livestock

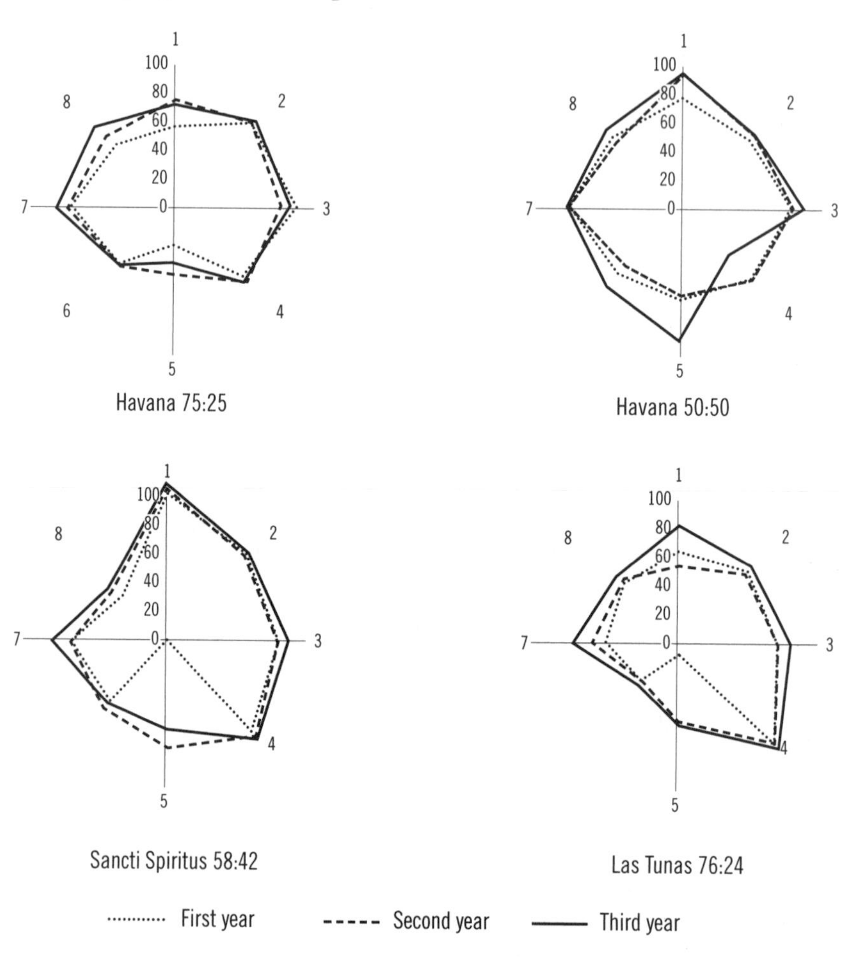

(Funes-Monzote and Monzote 2000)

Note: 1=reforestation, 2=total species, 3=number of food products, 4=labor intensity, 5=organic fertilizer production, 6=yield, 7=energy efficiency, 8=milk production

Research into integrated systems for agricultural production opens up a wide spectrum of possibilities for boosting farm viability and sustainability. In this type of research monitoring farms over time is critical to identifying the equilibria of system mechanisms and processes. Principle components analysis is one of the statistical methods successfully used in the evaluation and analysis of this kind of data. This method groups the variables studied into new variables that explain a high percentage of variations, showing significant correlation among some of the indicators. Cluster analysis has also been used to group farms and similar years, which can help identify stability over time.

The results of diversification of commercial dairy units—while falling well short of integration—show increases of milk yield and the total value of all agricultural production. Fewer losses were observed, and units became more profitable, with costs of 0.45 and 0.70 peso per peso of value produced.

Other sustainable livestock production systems, based on self-sufficiency in production of animal feeds under varying soil and climatic conditions, have demonstrated sufficient production of biomass to sustain milk yields with low external inputs. Milk yields/ha have reached 1,500 liters, with adequate animal health and reproduction. In the same systems, if a small amount of external inputs were purchased, such as protein meal, molasses, sugarcane, and concentrates (up to 0.5 t/animal/year), then up to 2,200 liters of milk/ha could be obtained. However, in the no external input systems the cost of one liter of milk was approximately 0.86 pesos, while it was 1.10 to 1.20 pesos in the latter systems, where the principle costs were in feed (Álvarez et al. 2000).

Conclusion

Despite significant production increases obtained over decades with the specialized dairy cattle production model, by the latter part of the 1980s the first signs of weakness in this model were detected. This was made abundantly clear with the drastic drop-off in the availability of inputs in the 1990s. The situation could have been even more critical if the Ministry of Agriculture had not taken measures to begin the transformation towards smaller-scale, diversified, and self-sufficient cattle production—which it could do based on research previously carried out.

Currently research toward a higher level of crop–livestock diversification and integration is underway. On-farm experience in cattle production is demonstrating the potential and viability of conversion to integrated crop/livestock systems. This concept has implications which go beyond the technological-productive sphere, directly or indirectly influencing the economic, social, and cultural conditions of farming families, by reinforcing their ability to sustain themselves. It is important to note the abundant knowledge and practical experiences that Cuban *campesinos* have already achieved with integrated systems over many years.

From the social and economic viewpoints, confronting labor shortages in rural areas is critical. The design of future agricultural systems is taking into account the need to stabilize rural populations and reverse rural-urban migration. This can be achieved by rearranging productive structures and investing in developing rural areas, giving farming more social and economical prestige.

The search for alternative livestock production systems is a top priority if we are to reach sustainability in the world's agriculture (Boehncke 1995).

It is the task of ecological animal and crop producers to demonstrate better, more integrated production systems that solve key problems in livestock production.

Agroecology provides the basic principles to study, design, and manage alternative agricultural ecosystems. It takes into account both the ecological and environmental dimension and the economic, social, and cultural aspects of the crisis of modern agriculture (Altieri 1995). The key agroecological strategy to achieve sustainable agricultural productivity is to change the monocultural structure of agriculture and reduce the dependence on external and imported inputs, by designing integrated agricultural ecosystems (Rosset 1998).

Cuba is taking important steps in the diversification and integration of crop and livestock production. These steps are backed by scientific research, and they will become the basis of conversion to organic, agroecological production. This will be a major asset for the Cuban economy.

References

Altieri, M. 1995. *Agroecology: The Science of Sustainable Agriculture.* Boulder, CO: Westview Press.

Álvarez, A., R. García-Vila, R. Ruiz, and H.J. Schwartz. 2000. Resultados del proyecto piloto lechero ACPA-Cuba sí. VII Congreso Panamericano de la Leche (FEPALE). Havana.

Anon. 1987. *La Revolución en la Agricultura.* Havana: José Martí, pg. 70.

Anon. 1998. Resolución Económica del V Congreso del Partido Comunista de Cuba. *Granma.* November 7.

Arteaga, O., M. Martínez, C. Hernández, and W. Espinosa. 1986. Determinación de la dosis técnico-económica de la aplicación de estiércol en suelos pardo grisáceos. *Resúmenes V Reunión ACPA.* Havana, pg. 56.

Boehncke, E. 1995. The future of organic livestock. *Ecology and Farming: IFOAM Journal.* September 14–18.

Borroto, A. 1988. Potencial forrajero de los subproductos agrícolas de cítricos para la producción de carne ovina (dissertation). Ciego de Ávila, Cuba: ISACA.

Castiñeiras, A., G. Jimeno, M. López, and L. Sosa. 1986. Control biológico de huevos de *Boopfhilus microplus* (Canestrini) (acarina: *ixodidae*) con hongos entomopatógenos y hormigas: Ensayo preliminar. *Resúmenes V Reunión ACPA.*

Crespo, G. 1984. El estiércol vacuno y su uso en la producción de los pastos. *Revista Cubana Ciencias Agrícolas* 18:249.

Crespo, G. and C. Hernández, 1985. Evaluación de la excreta de cerdo y gallinaza como fuente de fertilizante para los pastos. *Resúmenes del Evento Científico XX Aniversario del ICA.* Sección Pastos. Havana, pg. 18.

Funes, F. and H. Jordán. 1987. Producción y manejo: leche. *Leucaena, una opción para a la alimentación bovina en el trópico.* Havana: EDICA.

Funes, F. and M. Monzote. 1993. Pasture legumes in Cuba: Past, present, and future. *Proceedings of XVII International Grassland Congress.* New Zealand–Australia.

Funes, F., R. Pazos, and A. Álvarez. 1986. Elementos que indican la regionalización. *Resúmenes II Curso para directores de Empresas Pecuarias.* Area de Ganadería. Havana: MINAG, pg. 17.

Funes, F., S. Yepes, and D. Hernández. 1971a. Estudios de introducción de pastos en Cuba I. Principales gramíneas para corte, pastoreo y tierras bajas. *Memorias.* EEPF "Indio Hatuey": 17–39.

Funes, F., S. Yepes, and D. Hernández. 1971b. Estudios de introducción de pastos en Cuba II. Leguminosas más productivas. *Memorias.* EEPF "Indio Hatuey": 40–51.

Funes-Monzote, F. 1998. Sistemas de producción integrados ganadería-agricultura con bases agroecológicas. Análisis y situación perspectiva para la ganadería cubana (master's thesis). Spain: Universidad Internacional de Andalucía, pg. 43.

Funes-Monzote F. and D. Hernández. 1996. Algunas consideraciones y resultados sobre la elaboración y utilización del compost en fincas agroecológicas. *Agricultura Orgánica* 2:2.

Funes-Monzote F. and M. Monzote. 2000. Results on integrated crop-livestock-forestry systems with agroecological bases for the development of Cuban agriculture. 13th IFOAM International Scientific Conference. Basel, Switzerland.

Funes-Monzote, F., M. Monzote, D. Serrano, H.L. Martínez, and J. Fernández. 2000. Eficiencia energética y productiva de sistemas integrados ganadería-agricultura. I Congreso Internacional sobre Mejoramiento Animal. Havana: CIMA: 490–497.

García-López, R., E. Mora, A. Elías, R. García, and F. Alfonso. 1994. Evaluación comparativa de la saccharina húmeda (rústica) y la caña de azúcar fresca (con aditivos) para la producción de leche en secano. *Revista Cubana Ciencias Agrícolas* 28:1:47–50.

García Trujillo, R. 1993. Potencial y utilización de los pastos tropicales para la producción de leche. *Los Pastos en Cuba,* vol. II. Havana: EDICA, Instituto de Ciencia Animal.

García Trujillo, R. and M. Monzote. 1995. La ganadería cubana en una concepción agroecológica. Conferencias II Encuentro Nacional de Agricultura Orgánica: 60–68.

Hernández, N. 1989. Contribución al estudio de la regionalización de gramíneas en la provincia de Sancti Spritus (dissertation). Havana: ICA-ISCAH-MES.

Jeréz, I., M.A. Menchaca, and J.L. Rivero. 1986. Evaluación de tres gramíneas tropicales. II Efecto de la carga animal sobre la producción de leche. *Revista Cubana Ciencias Agrícolas* 20:231.

López, M. 1981. Las leguminosas tropicales de pastos y la simbiosis (dissertation). Havana: ICA-ISCAH-MES.

Marrero, D. 1989. Sistemas de alimentación con gramíneas y leguminosas para hembras de reemplazo (dissertation). Havana: ICA-ISCAH-MES.

Martín, P.C., E. Muñoz, J.B. Michelena, A. Zamora, D. Hernández, and N. Pérez. 1996. Resultados de UBPC, granjas y vaquerías con diferentes número de vacas en la provincia La Habana en 1994 y 1995 (internal report). Havana: ICA.

Martínez-Mojena, A. 1986. *Bacillus thuringiensis, Metarhizium anisopliae* y *Beauveria bassiana,* para el control de *Mocis latipes* y *Monecphora bicinta fraterna* en los pastos. VII Seminario Científico Nacional. Estación Experimental de Pastos y Forrajes "Indio Hatuey." Matanzas, Cuba.

Menéndez, J. 1982. Estudio y clasificación de las leguminosas forrajesras autóctonas y/o naturalizadas en Cuba (dissertation). Havana: ISCAH-MES.

MINAG. 1997. Legislación sobre las Unidades Básicas de Producción Cooperativas atendidas por el Ministerio de la Agricultura. Havana.

MINAG. 1998. Programa de alimentación para la seca 1998–1999 (internal document). Havana: MINAG.

Monzote, M. 1982. Mejoramiento de pastizales de gramíneas mediante la inclusión de leguminosas (dissertation). Havana: ICA-ISCAH-MES.

Monzote, M. and F. Funes-Monzote. 1997. Integración ganadería–agricultura. Una necesidad presente y futura. *Agricultura Orgánica* 3:1:7.

Monzote, M., F. Funes-Monzote, D. Serrano, J.J. Suárez, H.L. Martínez, J. Pereda, J. Fernández, A. González, M. Rodríguez, and L.A. Pérez-Olaya. 1999. Diseños para la Integración ganadería-agricultura a pequeña y mediana escala (final project report). Havana: CITMA.

Muñoz, E., G. Crespo, L. Fraga, and R. Ponce de León. 1993. Integración de la agricultura orgánica y la ganadería como vía de desarrollo sostenible de la producción de alimentos y protección del medio ambiente. Conferencias Primer Encuentro Nacional de Agricultura Orgánica. Havana: INCA-ISCAH, pp. 49–53.

ONE. 1995. Principales indicadores del sector agropecuario. Oficina Nacional de Estadísticas. August. Havana, pg. 30.

Pérez-Infante, F. and R. García-Vila. 1975. Uso de la caña de azúcar en la alimentación del ganado en la época seca. *Revista Cubana Ciencias Agrícolas* 9:2:109–112.

Pérez-Olaya, L.A. 1998. Regulación biótica de fitófagos en sistemas integrados de agricultura–ganadería (thesis). Havana: ISCAH.

Perón, E. and L. Marquez. 1992. Fincas Integrales. *Revista ACPA:* 2:13.

Rodríguez, M. 1998. Comportamiento de la Biota edáfica en sistemas integrados ganadería–agricultura (technical report). Havana: Proyecto CITMA.

Rodríguez, S. 1998. Freno al decrecimiento de la ganadería. *Negocios en Cuba, Cuba en cifras y hechos.* Havana.

Rosset, P.M. 1998. La crisis de la agricultura convencional, la sustitución de insumos, y el enfoque agroecológico. *Food First Policy Brief* No. 3. Oakland, CA: Institute for Food and Development Policy.

Simón, L. and M. Esperance. 1997. El silvopastoreo, una alternativa para mejorar la eficiencia del uso de la tierra en los cítricos. *Agricultura Orgánica* 3:1:14–15.

Sosa, M. and Funes-Monzote, F. 1998. *Energía: Sistema computarizado para el análisis de la eficiencia energética.* Havana: IIPF-ONP.

Stuart, R., A. Elias, A. Delgado, and E. Muñoz. 1985. La caña de azúcar y sus subproductos en la alimentación de los rumiantes (roundtable report). Evento Científico XX Aniversario. Havana: ICA.

Tang, M. 1986. Inoculación de cepas de *Rhizobium* en leguminosas tropicales. *Resúmenes V Reunión ACPA.* Havana, pg. 51.

Ugarte, J. 1972. Amamantamiento restringido. *Revista Cubana Ciencias Agrícolas* 6:185.

PART III
Examples and Case Studies

CHAPTER TWELVE

Green Medicine: An Option of Richness

Mercedes García, Pasture and Forage Research Institute

"Green medicine is not a compromise with poverty, but an option of richness."

—General Raúl Castro, Minister of Defense

Green medicine, so-called because its many products are prepared from medicinal plants, has been transmitted from generation to generation over millennia, and is still in vogue worldwide. The multiracial character—a mixture of Spaniards, Africans, Chinese, and other Europeans and Asians, with some influence of the mostly decimated indigenous peoples—yet unified national identity of Cubans has contributed to a rich diversity and knowledge of medicinal plants.

In Cuba the use of medicinal plants was never abandoned, as even the many who moved from the countryside to the city have retained confidence in the curative powers of green medicine. Despite the boom of "industrial" medicine and the growth of the public health system over the past 40 years, the use of medicinal plants continued in popular culture. Thus medicinal plants have always been important to the great majority of the population, and when shortages of industrial medicines at the beginning of the "Special Period" led to a national reassessment of green medicines, this was welcomed by the population. Thankfully, ample scientific information was already available.

The esteemed Cuban scientist, Dr. Juan Tomás Roig y Mesa—considered the father of Cuban green medicine—devoted years to the study of Cuban flora (Roig 1942a, 1942b, 1945a, 1945b). In 1973 the Medicinal Plants Experiment Station was established in Havana province. Its researchers initially followed up on Dr. Roig's botanical work, building a collection of more than 5,000 species. Their results were published as *Medicinal, Aromatic, and Poisonous Plants of Cuba* (*Plantas Medicinales, Aromáticas y Venenosas de Cuba,* Roig 1974). As early as 1982 the *Medicinal Plants Bulletin* (*Plantas Medicinales: Boletín de Reseñas*) began to publish research reports on Cuban medicinal flora (Lima 1988; Fuentes et al. 1988, 1989; Fuentes and Granda 1989).

Today in Cuba there is a new, large, complex research and production system for medicinal plants. In 1988 fifty-two medicinal extracts were imported, 25 percent of which could possibly have been produced in the country (Fuentes 1988). Now some 60 extracts are regularly produced in Cuba. Medicinal plants are processed into many medicinal products sold in pharmacies across the country, contributing immeasurably to the health of the population and saving the country vast quantities of foreign exchange.

How Was It Possible to Build a System in Such a Short Time?

In order to understand how Cuba could rapidly build up a green medicine system, we need to look at the conditions faced by the island over many years. Because of the constant hostility of the United States, the Revolutionary Armed Forces (FAR) had long envisioned as part of their wartime plans that the country would be totally blockaded, unable to receive any external supplies. Under such circumstances, Cuba would have to manufacture medicines from plants, using minimal resources, with sufficient quality control to guarantee the health of the troops and the population in general. Military policy saw it as a strategic issue that the country be prepared to be self-sufficient in medicines in the event of a total blockade, so the Ministry of Defense (MINFAR) opened the Central Laboratory of Herbal Medicine at the Higher Institute of Military Medicine.

The economic situation at the beginning of the 1990s, after the collapse of trade with the former Socialist Bloc—which had been the main source of inputs and raw materials for industrial pharmaceuticals—differed only slightly from the state of war conceived years earlier by MINFAR. By the time the crisis came, the army had already gathered all the necessary information from research carried out over many years and knew how to begin work at once. Also key was the tremendous number of scientifically trained people in Cuba. With only two percent of the population of Latin America, Cuba has 11 percent of its scientists and technicians, as well as a substantial research infrastructure, which made possible the rapid replacement of no-longer available products and inputs with Cuban technological innovations.

The first step was to determine which plant species had the most potential based on both proven therapeutic properties and agronomic characteristics. This task was entrusted to a national group of experts from the Ministries of Public Health (MINSAP), Agriculture (MINAG), Science, Technology, and the Environment (CITMA), and MINFAR. Coordination between MINSAP and MINAG led to the rapid authorization of the medicinal plants that would initially be used by the national health care system. In 1990 MINAG created the Department of Medicinal Plants to oversee the production of selected species in sufficient quantity.

Medical trials were quickly replicated in different provinces of the country, to evaluate curative properties—especially for the elderly—and statistical comparisons were made in the countryside and the cities (Acosta 1993). Another group of researchers from various institutions set up agronomic trials on one hundred species in two regions that differ climactically and geographically. Forty of the species are normally found in both zones. Their agronomy was studied for more than three years to determine the impact of climate and altitude on production and quality. Plants were classified based on flowering time, season, and other variables, followed by research on the translocation within the plant of the chemical compounds with medicinal properties. These results provided crucial information as to optimal planting and harvesting dates—even to the best hour of the day—and which plant parts to harvest in order to obtain the best quantity and quality of the desired chemical compounds. From this work some thirty species were selected for the treatment of the most common diseases in Cuba, and they were evaluated and compared to medicinal uses reported in other countries before being recommended for clinical use. This research is ongoing.

In 1995 MINAG sponsored the First National Technical Congress on the Agronomy and Marketing of Medicinal Plants, where more than 100 experiments were reported. This included trials carried out by researchers, technicians, and peasant farmers. The diversity of topics and the high level of interest expressed by the participants was a testimony to what had already been achieved, and generated a great deal of enthusiasm for new projects.

Fresh medicinal plants are now grown in orchards, courtyards, and organoponic urban gardens, and are marketed directly to people who prefer to prepare herbal infusions at home. These types of relatively informal production must still meet stringent quality control requirements set by the state.

A decision was made early on to set up specialized farms for growing medicinal plants in every province. The assigned task of these farms is to grow in bulk the plants approved by the national body, following official standards in terms of sowing date and norms, agronomic practices, harvesting date, and those plant parts with the most concentrated active agents. These specialized farms guarantee a high degree of quality control of both fresh and dry products, though most of the green medicine from these farms is sold dry.

Production is based on annual and monthly plans, developed in coordination with MINSAP to take into account information on disease incidence at the municipal and provincial level. This basic information comes from family physicians and natural medicine clinics. By 1993, these medicinal plant farms had 343 hectares producing 261.6 tons of green matter (Pagés 1993).

These farms do not rely on monocrop methods; intercropping of medicinal plants, association with food crops, and crop rotations are all commonly practiced. It is critical that the agronomic process used to produce medicinal plants be essentially organic. Chemical fertilizers and pesticides cannot be used on such plants, since they can affect human health.

At the María Teresa farm, for example, located in Havana province, 75 species of medicinal plants out of a total of 120–150 crop species are grown on 65 hectares under strict organic farming conditions. The farm uses vermiculture to produce its own high-quality fertilizer of worm castings, and produces its own seed supply. This is the most important medicinal plant farm in the country as it covers the needs of Havana City and Havana province, where the greatest proportion of the Cuban population is found. Production is based on plans that take into consideration the species, quantities, and delivery dates of the product to be turned over to the processing centers. Special areas are set aside to supply the pharmaceutical industry with the ingredients needed to prepare the more sophisticated products.

Harvesting chamomile (Matricaria chamomilla) *by hand.*

Production of green medicines at the Pasture and Forage Research Institute in Havana province.

Special processing centers have been set up in each zone, where dry plant material is received along with other inputs (natural alcohol, sugar, etc.). These centers make tinctures and liquid extracts, which may be sold directly as is, or may serve as active ingredients for syrups, elixirs, ointments, or cream preparations. One special line of products, called "Api-pharmaceuticals" (from apiculture, or beekeeping), incorporates honey along with medicinal plant tinctures or extracts, and is sold under the name "Melitos" (from *miel*, the Spanish word for honey). The honey adds curative properties to the plant mixture (see Table 1). The commercial Biological Pharmaceuticals Laboratory (LABIOFAM) manufactures veterinary medicines of plant origin, and has acquired international prestige due to the very effective products it produces.

An educational and training program on natural and traditional medicine is being developed. Reference materials have been prepared for physicians, such as the *Therapeutic Guide to Dispensing Green Medicines* (*Guía Terapéutica Dispensarial de Fitofármacos y Apifármacos,* MINSAP 1992), the series of books *FITOMED I, II,* and *III* (MINSAP 1993), and other manuals for prescribing natural medicines. Such books usually have the botanical description of each plant with information that facilitates its identification, identifies therapeutic properties, procedures for preparation, recommended dosages, and relevant warnings, if there are any.

Table 1. Medicinal plants commonly used in Cuba and the ailments they treat

COMMON NAME	SCIENTIFIC NAME	AILMENT
Chili pepper	*Capsicum annuum*	Rheumatism
Garlic	*Allium sativum*	Asthma, common cold, circulatory ailments, stomach ailments, fungal infections, parasites, high blood pressure, back pain
White basil	*Ocimum basilicum*	Stomach ailments, high blood pressure
Wild indigo	*Indigofera suffruticosa*	Lice
Anis	*Piper auritum*	Stomach and joint aches, rheumatism
Gourd squash	*Cucurbita moschata*	Parasites
Lemongrass	*Cymbopogon citratus*	Asthma, common cold, fungal infections, high blood pressure, throat ailments
Senna	*Cassia grandis*	Fungal infections
Dill	*Anethum graveolens*	Stomach ache
Eucalyptus	*Eucalyptus globulus*	Asthma, common cold, earache, cough
Lemon eucalyptus	*Eucalyptus citriodora*	Fever, cough
French senna	*Cassia alata*	Fungal infections
Guava	*Psidium guajava*	Diarrhea, fungal infections
Mint	*Mentha spicata*	Common cold, gastritis
Fennel	*Foeniculum vulgare*	Gastritis
Royal ítamo	*Pedilanthus tithymaloides*	Stomach ailments, mouth sores
Five-leaf jasmine	*Jasminum grandifolium*	Nervous disorders
Ginger	*Zingiber officinale*	Gastritis, cough, vomiting
Lemon	*Citrus aurantifolia*	Circulatory ailments, stomach ailments
Plantago	*Plantago major*	Common cold, stomach ailments, mouth sores, burns, cough
Chamomile	*Matricaria chamomilla*	Diarrhea, stomach ache, fungal infections, gastritis, mouth sores
Japanese mint	*Mentha arvensis*	Stomach ache, gastritis
Muraya	*Murraya paniculata*	Headache
Sour orange	*Citrus aurantium*	Circulatory ailments, stomach ache
French oregano	*Coleus amboinicus*	Common cold, cough
Passion fruit	*Passiflora incarnata*	Nervous disorders
Licorice verbena	*Lippia alba*	Headache
Rue	*Ruta graveolens*	Nervous disorders
Aloe	*Aloe barbadensis*	Asthma, common cold, minor cuts and bruises, burns
Sago palm	*Maranta arundinacea*	Diarrhea
Sage	*Salvia officinalis*	Common cold, fungal infections, kidney ailments
Tamarind	*Tamarindus indica*	Constipation, kidney ailments
Kidney tea	*Orthosipon aristatus*	Kidney ailments
Linden	*Justicia pectoralis*	Common cold, nervous disorders
Orange mint	*Mentha citrata*	Diarrhea, fever
Periwinkle	*Lochnera rosea*	Conjunctivitis

The Pasture and Forage Research Institute published a very popular book in 1995 titled *Identifying and Using Medicinal Plants*. Families have found it very useful for home use. It contains most of the same information as the technical books, but in language making it more accessible (García 1995). The book has illustrations of each plant, and agronomic information, so that anyone with a small plot, garden, or even flower pots can raise their own medicinal plants. Other popular materials have been published by other institutions, such as *The Family Book* (*El Libro de la Familia*) issued by the FAR in 1991, and many short courses and workshops have been offered in neighborhoods and communities promoting green medicine use at home.

Advertising campaigns are employed to promote green medicine using the print media, radio, and TV. The *Trabajadores* and *Juventud Rebelde* national newspapers, as well as some provincial papers, regularly devote columns to the use of medicinal plants. Informational spots are also broadcast on television.

In recent years many natural and traditional medicine clinics have been opened, where primary health care services are provided. They offer not only green medicine, but also other therapies that take a holistic approach. These clinics have specialists in natural medicine and acupuncture, acupressure, physical therapy, Chinese medicine, and other techniques. Today there are clinics of this type in all provinces and in most municipalities, and their reputation is growing thanks to the quality of their services, most notably for the elderly and those suffering from hypertension, diabetes, and other illnesses for which it is wise to avoid the use of chemical medicines.

Conclusion

In an innovative mixture of tradition and science, the Cuban government has incorporated the use of medicinal plants in an internationally recognized health system. The Ministries of Agriculture (MINAG) and Health (MINSAP) have closely collaborated in successful research for the production and marketing of green medicines on a large scale since 1990, taking advantage of traditional knowledge, enriched by scientific research. The rapid success achieved in the past decade was made possible by earlier comprehensive studies of Cuban medicinal flora and a solid health care structure. Cuban citizens were predisposed in favor of green medicine, and responded positively to government initiatives in this direction during the difficult "Special Period." Training programs for health care personnel and the general public, in workshops and by means of the mass media, have helped. Foreign experts in natural medicines have also contributed. Current plans call for a deepening and broadening of the use and knowledge of medicinal plants in Cuba.

References

Acosta, L. 1993. Proporciónese salud. Cultive plantas medicinales. Havana: Ediciones Científico-Técnicas.

FAR. 1991. *El Libro de la Familia.* Colección Verde Olivo. Havana: MINFAR.

Fuentes, V.R. 1988. Las Plantas Medicinales en Cuba (dissertation). Havana.

Fuentes, V.R. and M.M. Granda. 1989. Potencialidad fitoquímica de la flora de Cuba. *Plantas Medicinales. Boletín de Reseñas* 20.

Fuentes, V.R., D.E. Ordaz, and M.M. Granda. 1989. Comparación de la utilización de las plantas medicinales en la medicina tradicional de varios paises. *Plantas Medicinales. Boletín de Reseñas* 20.

Fuentes, V.R., N. Rodríguez, and D.E. Ordaz. 1988. Plantas medicinales de uso popular referidas como tóxicas. *Plantas Medicinales. Boletín de Reseñas* 19.

García, M. 1995. *Saber y hacer sobre plantas medicinales.* Havana: IIPF-CIC. Ediciones Programa Biovida.

Lima, H. 1988. Sobre el mejoramiento genético de plantas medicinales. *Revista Cubana de Farmacéuticos* 22:3:113–118.

MINSAP. 1992. *Guía Terapéutica Dispensarial de Fitofármacos y Apifármacos.* Havana: Minsisterio de Salud Pública.

MINSAP. 1993. *Plantas Medicinales. FITOMED I, II, and III.* Havana: Ministerio de Salud Pública.

Pagés, R. 1993. Avanza programa de apoyo a la industria farmacéutica. *Granma,* August 19.

Roig, J.T. 1942a. El cultivo y la industria de plantas medicinales en Cuba. *Almanaque Agrícola Nacional* 196–200.

Roig, J.T. 1942b. Cultivo de plantas medicinales en Cuba. *Revista Nacional* 1:4.

Roig, J.T. 1945a. Plantas Medicinales y Aromáticas. *Revista de Agricultura.* Havana: MINAG.

Roig, J.T. 1945b. *Plantas Medicinales, Aromáticas o Venenosas de Cuba.* Havana: Primera Edición.

Roig, J.T. 1974. *Plantas Medicinales, Aromáticas o Venenosas de Cuba* (third edition). Academia de Ciencias de Cuba. Ediciones Científico-Técnicas. Havana: Instituto del Libro.

CHAPTER THIRTEEN

The Growth of Urban Agriculture

Nelso Companioni and Yanet Ojeda Hernández, National Institute for Fundamental Research on Tropical Agriculture, Egidio Páez, Cuban Association of Agricultural and Forestry Technicians, City of Havana Chapter, and Catherine Murphy, Food First/Institute for Food and Development Policy

In recent years a strong urban agriculture movement has developed in Cuban cities and suburbs.[1] The goal of this movement is to maximize the production of diverse, fresh, and safe crops from every patch of previously unused urban land. This urban production is based on three principles: organic methods, which do not contaminate the environment; the rational use of local resources; and the direct marketing of produce to consumers. Thus our urban agriculture fits the concept of sustainability, which in this context means the abundant use of organic matter and biological pest controls, and adherence to the principle of local inputs. The local sale of produce has played a significant role in meeting local food requirements (Companioni et al. 1997).

Urban agriculture has characteristics that differentiate it from conventional agriculture and large-scale production systems, such as the diversity of production and the many people who participate. This gives a special aspect to extension work, where new management models and work styles must be developed to achieve sustainable production levels in each neighborhood and region.

Urban agriculture is participatory, not only in the sense of the broad base of involvement, but also in demanding diverse responses to the heterogeneity of local conditions, requiring a variety of techniques to create the best possible conditions for production. Because of its geographical location and intended market, this movement must be low-input, use no toxic pesticides, make efficient use of water, and carefully manage soil fertility and the cultivation of its crops and animals. Urban agriculture has received and receives special attention from the highest levels of the Ministry of Agriculture and other government officials.

Urban Agriculture Yesterday and Today

During the first half of the century, urban agriculture existed on a small scale, involving a few people, and was aimed at the production of a few plant species (mainly leafy vegetables), the rearing of domestic animals in backyards, and provisioning of foods for families.

Beginning in the 1960s, high-tech horticultural production techniques were introduced, based on complex technology and crop management systems with a heavy use of chemical products, as are found in hydroponics and "zeoponics" (production in zeolite susbtrate). The high degree of specialization in these production systems and the development of large enterprises for producing vegetables and other crops during the 1970s and 1980s monopolized the vegetable market, relegating small-scale producers to a second tier (Companioni et al. 1996a).

Like many other countries, Cuba after the Revolution opted to be in the vanguard of the "Green Revolution," which involved the industrialization of agriculture and the adoption of practices aimed at producing sufficient food for the country. After the Earth Summit in Río, Cuba moved with the world community towards a new emphasis on a more natural agricultural system, through which food quality, the nutrition of the population, and natural resources would benefit. Urban agriculture reemerged recently in this new context for several reasons: the economic difficulties of the 1990s; the low quality of vegetables on the market; shortages of traditional spices and seasonings; and the under-exploited production potential of cities. With a renewed emphasis on urban farming, relatively high levels of production in small areas were made possible by paying close attention to existing local resources, and the potential for selling goods locally.

The simplicity of this form of production, and the increase in yields while still improving the technology, allowed for the rapid development of the popular movement in urban agriculture. This new sector has created 160,000 jobs, taken by people of various occupations and backgrounds, including workers, masons, mechanics, housewives, retired people, and professionals (López 2000).

Employing a large number of people in urban agriculture is one of the greatest social impacts of this movement. Driving these changes was the potential for increased income generated by selling produce from urban gardens; this attracted the attention not only of workers, but also professionals from diverse backgrounds, who received state supports in the form of land, credit,

Urban agriculture: public sales outlet, Havana.

services, and/or inputs. This new agricultural labor force has brought dynamism and innovation to every municipality.

In each territorial unit, services for urban agriculture are grouped together in what we call a Municipal Urban Farm Enterprise. This administrative unit helps coordinate all urban agricultural activities in the municipality in a variety of ways. It is the source for extension and technical assistance, helps link urban farmers and gardeners with each other, and builds links with research, educational, and service centers (Ojeda 1999). Each Municipal Urban Farm also has the responsibility of organizing production and determining the appropriateness of different technologies for each of its subunits, taking into account local resources, inputs, and the potential of the land. Intense technical training for producers has played a decisive role in the achievements attained so far.

Premises of Urban Agriculture

Several basic premises explain the strong potential of urban agriculture:

- Urban centers have the highest demand for those foodstuffs which are easily perishable when transported. Thus there is a basic logic to the notion that perishable foods should be produced as near as possible to the consumer.

- Vegetables, fruits, flowers, spices, and intensive animal production all require a large labor force, which is available in towns and cities. In Cuba, 75 percent of the population is urban, most of which came originally from rural areas; thus, many urban dwellers have empirical knowledge about crop management and livestock production.
- The growth and spread of cities invariably creates many empty spaces in peripherial areas, which often become trash-dumps that are sources of disease vectors, are a danger to human health, and despoil the urban environment. Using such areas for food production has eliminated these dangers and has made Cuba's cities healthier and more beautiful.

Basic Principles

Planning for urban agriculture in Cuba is guided by a set of basic principles defining its objectives and organization. Among them are the following:

- uniform distribution throughout the country
- logical correspondence between production and the number of dwellers of each region
- crop-animal integration with maximum use of synergies to boost the production of each
- intensive use of organic matter to boost and preserve soil fertility, and biological pest controls
- use of each patch of available land to produce food, guaranteeing intensive production and high yields of crops and animals
- multidisciplinary integration and the intense application of science and technology
- a fresh supply of good quality products, offered directly to the population, guaranteeing a balanced production of not less than 300g of vegetables daily per capita and an adequate variety of animal protein sources
- maximum use of the potential to produce food, such as the labor force available and the recycling of wastes and by-products for plant and animal nutrition

Organizational Structure of Urban Agriculture in Cuba

In Cuba today there is an urban agriculture structure in all cities and towns, thanks to the impact and rapid development of this popular form of food production and to the degree of urbanization of the Cuban population (see Table 1).

Table 1. Urbanization in Cuban provinces

PROVINCE	% URBAN
Pinar del Río	63.9
Havana	78.4
City of Havana	100.0
Matanzas	80.3
Villa Clara	77.5
Cienfuegos	80.7
Sancti Spíritus	69.7
Ciego de Ávila	74.6
Camagüey	75.0
Las Tunas	58.8
Holguín	59.0
Granma	57.6
Santiago de Cuba	70.2
Guantánamo	59.6

The National Urban Agriculture Group composed of specialists and government officials from different scientific and government institutions, and urban farmers—regulates and directs this effort, exercising its influence at different levels all to way to the grassroots through provincial and municipal groups. Regional and local groups are responsible for the organization, development, and regulation of urban agriculture in their zone, and the coordination between all entities and persons related to production, processing, and distribution of food within the boundaries of each territory and province.

Within each Popular Council (local government at the neighborhood level), a representative or agricultural delegate coordinates urban agriculture. Likewise, many activities related to urban agriculture—such as veterinary medicine, plant protection, and biopesticide production—are represented at the Popular Council. Different areas of responsibility are coordinated through the Popular Councils, which take into consideration the unique characteristics of local systems of production and oversee technical and service units such as the veterinary clinic, farmers' shop, nurseries, laboratories for the production of biological products, and others.

Within a municipality, the coordinating activities of the Popular Councils are carried out through the Municipal Urban Farm, which in addition to its coordinating role, has the infrastructure necessary to carry out technical and service activities, with the capability to gather together scientific and technical resources and farmers from different productive areas and related institutions within its territory.

Twenty-six administrative sub-programs attend to urban agriculture. These are tied to specific topics such as vegetable production, medicinal plants, herbs, grains, fruits, and rearing of animals (hens, rabbits, sheep, goats, pigs, bees, and fish), all of which can be found throughout the country (Table 2).

Table 2. Current sub-programs of Cuban urban agriculture

1. Soil management and conservation	14. Oilseed crops
2. Organic matter	15. Beans
3. Seeds	16. Animal feeds
4. Irrigation and drainage	17. Apiculture
5. Vegetables and fresh herbs	18. Poultry
6. Medicinal plants and dried herbs	19. Rabbit breeding
7. Ornamental plants and flowers	20. Sheep and goats
8. Fruit trees	21. Swine
9. Shade houses	22. Cows
10. Small-scale "popular" rice production	23. Aquaculture
11. Trees, coffee, and cocoa	24. Marketing
12. Small-scale "popular" plantain production	25. Small-scale agro-industry
13. Tropical roots and tubers	26. Science, technology, training, and environmental issues

(GNAU 2000)

VEGETABLES AND FRESH HERBS (ORGANOPONICS, INTENSIVE VEGETABLE GARDENS, SMALL PLOTS, AND BACKYARDS)

This was the first urban agriculture activity carried out, and hence is the most developed. The goal for this type of urban agriculture has been set at producing 30 million quintals[2] (1,380,000 metric tons) of fresh vegetables per year, with yields above 20 kg/m^2 per year in organoponics (raised beds filled with organic matter), 10 kg/m^2 per year in intensive vegetable gardens, and 10 kg/m^2 per year in small plots and backyards. The goal for the end of 2000 was to have at least 5 m^2 per inhabitant dedicated to these types of production, making a substantial contribution to the national goal for all vegetable production of 300 g of fresh vegetables daily per capita.

Organoponic intensive farming at the "Gilberto León" Cooperative.

The heterogeneity of Cuba and the diversity of possible ways to grow vegetables have combined to generate distinct production systems. The most common are the following:

ORGANOPONICS AND INTENSIVE VEGETABLE GARDENING: These have been the most important methods over the past years, and have gone a long way toward helping us rediscover our horticultural traditions. These systems are an example of how scientists and gardeners can work together to develop new production methods (MINAG 2000). The main difference between these two systems of production lies in the fact that organoponics are generally located in areas with infertile soils or with production constraints. For example, organoponics can be built on artificial surfaces, on which containers are placed and filled with a mixture of organic matter substrate and soil, in which to grow the crops. The intensive vegetable garden is developed on parcels of relatively good soil, without using raised beds, though organic matter is applied directly during preparation for planting (Peña 1995, 1998).

SMALL PLOTS, PATIOS, AND POPULAR GARDENS: In this popular form of production, as a rule, the area cultivated is very small and is determined by how much useful or arable space exists between buildings, between houses and streets, or in a patio, or a state-owned urban space that can be converted to gardens. In general, the small plots, patios, and popular gardens situated in suburban areas are larger than those in the city centers. This type of production now makes significant contributions to household and regional food supplies. At this point there are 104,087 parcels and patios under production, covering an area of 3,595 hectares, which produce more than organoponics and intensive gardens combined. This type of land use has several positive effects. The small plots, patios, and popular gardens have made it possible to feed the urban population; have promoted the development of an urban culture favorable to agriculture; have eliminated the abandoned spaces which in the past may have been breeding grounds for disease vectors and rodents; and have provided socially useful and productive employment opportunities (Ojeda 1997).

SELF-PROVISIONING AT FACTORIES, OFFICES, AND BUSINESSES: The concentration of industrial production and of innumerable educational, health, and service facilities in the main population centers demands the operation of hundreds of worker's cafeterias, whose food supply requires large quantities of diverse agricultural products. A considerable number of workplaces have organized agricultural production in areas bordering, or close to, their facilities. This helps to meet the cafeterias' demand for food without reducing the food resources available to others in the neighborhood. The magnitude of this form of production has reached a level such that it must be considered a separate form of urban agriculture, particularly because of the differences between these self-provisioning farms and others. In the country's capital alone, there are more than 300 such farms in production. They total an area of 5,368 hectares, on which large quantities of vegetables, root crops, grains, and fruits, as well as meat, milk, fish, eggs, and herbs are produced.

SUBURBAN FARMS: On the edge or outer ring of Cuban cities we find many integrated suburban farms. They grow from the urbanism movement, are considered part of the urban environment, and are a key feature in planning for urban growth and development. Although they could never meet all the food needs of the urban population, they are larger than, and have a higher degree of integration among their sub-components than do the small plots and intensive gardens in the interior of the cities. Typically they cover between 2 and 15 hectares. The type of production found on a suburban farm will be strongly influenced by the surrounding population. We can see this from the point of view of infrastructure, recycling of waste products, the crops grown and animals raised, how the products are marketed, etc. This form of agricultural production is characterized by intensive cultivation, efficiency of water use, and the maximum reduction of agrotoxins. Suburban farms have reached an important level in the past few years, particularly in the cities of Havana, Santa Clara, Sancti Spíritus, Camagüey, and Santiago de Cuba. In the city of Havana, 2,000 small private farms and 285 state farms are under production; together these cover an area of 7,718 hectares, and are highly productive.

SHADED CULTIVATION AND APARTMENT-STYLE PRODUCTION: These systems are in their initial phases of development. Covered or shaded production utilizes mesh tents or "shade houses" of Spanish, Israeli, or Cuban design for growing crops and germinating seedlings. Work is underway to develop a complete technology appropriate for Cuban conditions, allowing for the year-round cultivation of horticultural crops, especially during the hottest months when the sun is at its most intense. Apartment-style agriculture is very diverse. It includes a range of practices, including cultivation with diverse soil substrates and nutrient solutions, mini-planting beds, small containers, balconies, roofs, etc., with minimal use of soil. This type of production has its own unique technologies and forms of organization (Carrión 1996).

RESULTS OF THE VEGETABLE SUB-PROGRAM

In recent years this program has experienced sustainable growth, both in production levels and in the yields obtained (see Figure 1). During 1999 vegetable production in organoponics and intensive gardens provided the population with 215 g/day per person of fresh horticultural crops (MINAG 1999), which represents a little more than half of the goals set (see Table 3). The most success has been achieved in Cienfuegos, Ciego de Ávila, Sancti Spíritus, and Havana. This program has been the laboratory for testing, confirming, and consolidating the principles, objectives, and overall perspectives for urban agriculture in Cuba.

Figure 1. Vegetable production from organoponics and intensive vegetable gardens

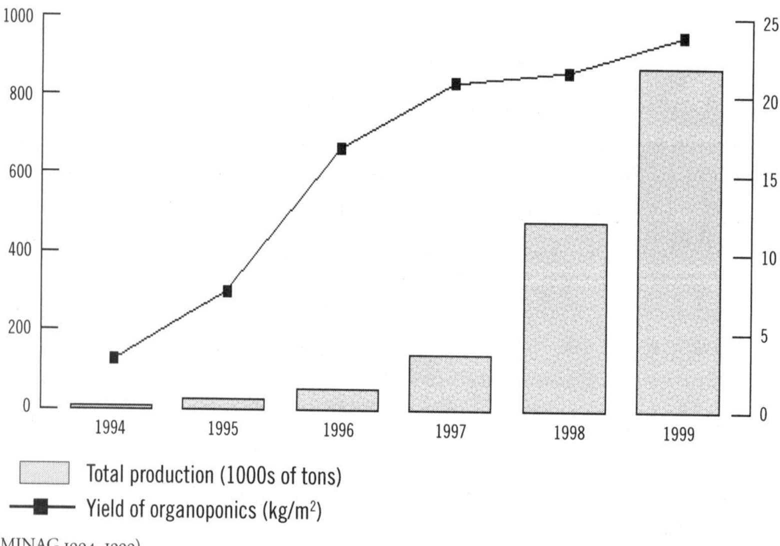

(MINAG 1994–1999)

Table 3. National vegetable production from organoponics and intensive gardens, 1999

PROVINCE	POPULATION	AREA (HA)	PRODUCTION	G/DAY/PERSON
Pinar del Río	726,929	602	73.0	274
Havana	689,364	712	88.9	351
City of Havana	2,197,706	462	70.2	88
Matanzas	649,994	382	59.2	249
Villa Clara	830,085	504	65.7	216
Cienfuegos	389,541	402	63.3	442
Sancti Spíritus	456,294	457	60.9	368
Ciego de Ávila	400,720	473	58.8	399
Camagüey	778,772	312	76.6	269
Las Tunas	521,793	314	36.9	193
Holguín	1,018, 899	663	58.3	153
Granma	823,481	366	56.1	186
Santiago de Cuba	1,022,105	398	47.9	128
Guantánamo	509,210	162	55.6	299
Isle of Youth	78,259	31	4.6	162
Total	11,093,152	6,213	876.0	215

SMALL-SCALE "POPULAR" RICE PRODUCTION

This sub-program has made significant advances in the past three years. Small-scale rice production is growing in all of the country's provinces, and the use of local resources in crop management has generated rice yields above 5 tons/ha in many units, a higher level than that which is achieved on state farms.

MEDICINAL PLANTS AND DRIED HERBS

Like the previous program, this is a recent one within urban agriculture. Herbs and medicinal plants are grown in organoponics and intensive vegetable gardens; yet they have their own program, which means their production is taken into account in regional planning according to local needs. In some cases, a portion of the production is sold via the Ministry of Public Health for processing into "green" medicines, which are distributed through the network of public pharmacies. The rest is sold fresh or dry for domestic consumption. Dried herbs are destined for Cuban kitchens; high levels of production make drying essential. Consumption of dried seasonings in Cuba has reached an annual per capita level of 120 g. An intensive educational and promotional campaign has been carried out to promote knowledge of their preservation, processing, and home use, through publications and radio and TV programs (Figueroa and Lama 1997, 1998, and 1999).

Ornamental plant nursery at the Alamar UBPC in Havana.

ORNAMENTAL PLANTS AND FLOWERS

This is the least advanced sub-program in the majority of the provinces. There are only a few units dedicated specifically to flower production. This program has grown in areas around Havana and also others such as Camagüey and Ciego de Ávila. The initial goal is to produce five dozen flowers per capita per year.

FRUIT TREES

Despite being a recent addition to the urban agriculture movement, the planting, care, and uses of a variety of fruit trees along urban perimeters has long been a tradition in Cuba. This sub-program has demonstrated a high productive potential, especially in mangoes, avocados, and citrus. Current plans contemplate a broad program of nurseries and grafting in future years in order to accelerate urban fruit production.

POULTRY

This sub-program, specializing in hens and ducks, is the most advanced of the animal production programs in urban agriculture. Producers are assigned ten females and one male of the semi-rustic local chicken breed. This breed has been produced by crossing a locally adapted creole hen with a more productive line of hen, such as a Rhode Island Red. From this cross birds were obtained that are characterized by their resistance to environmental adversity and high productivity of meat and eggs. During their adult stages, this hen, with good feeding (109 g/bird/day), will lay eggs year-round with an average annual production of 200 eggs/bird.

A certain amount of progress has been reached with rearing ducks, as it is the domesticated bird with the fastest growth rate. In just seven to eight weeks, ducks can reach between 2.8 and 3.2 kg (live weight), converting close to three kg of feed for each kg of weight gained. Ducks are also less sensitive to environmental stress and food quality, and more resistant to some infectious diseases that are common in birds. In addition to chickens and ducks, geese, turkeys, and guinea hens are also produced on a small scale (Companioni et al. 1996b).

SWINE

This sub-program has special features because rearing pigs within city boundaries requires strict sanitary control measures and vaccination. This program is focused in suburban areas, under the following requirements defined by the Institute of Veterinary Medicine (IMV):

- adequate food supply
- sufficient water supply for drinking and hygiene
- confinement
- a residue pit or biogas digester
- a cement or tile floor, and a roof for protection from weather

To begin rearing pigs, the prospective producer must sign agreements with the swine production group and the Territorial Technical Service for Swine Production. Through these agreements the producer purchases 12–20 kg piglets at a reduced price, as well as part of the necessary feed for fattening. After four to five months, when the pig reaches 90 kg or more, the contracted quantity of meat agreed upon by the producer is purchased by the state at the official price, and the surplus is sold at higher prices.

If the new pig farmer can produce or find his own pig feed, he need only buy a vitamin and mineral supplement for his animals. To fatten 40 animals on a 140-day cycle, and to finish 100 animals in a year at an average weight of 90 kg, it would be necessary to plant 4 hectares of soybeans, 7 hectares of sunflowers, and 6 hectares of sugarcane.

ORGANIC MATTER

Among the working guidelines for urban agriculture is "to systematically apply organic matter by using all available local alternatives, and to systematically develop local programs to assure adequate supplies of organic matter." In view of the importance of this activity, and to realize its potential, a sub-program was created in charge of organization, promotion and development of organic matter sources, and to assure their collection, processing, conservation, and distribution (GNAU 2000). The National Urban Agriculture Group (GNAU) coordinates these activities, supported by the Organic Fertilizers Reference Center, located at the National Institute for Fundamental Research on Tropical Agriculture (INIFAT) in the City of Havana, as well as by provincial and municipal organic fertilizer centers. This structure extends to the grassroots with centers organized by each Popular Council which receive technical assistance from a Technical Operations Group made up of specialists and farmers from different organizations and institutions. Territorial Organic Fertilizers Centers are responsible for organizing and advising the activities in their territory, geared toward the largest possible proliferation of small production units located at the sources of organic matter or at agricultural production units, to get this important input directly to the farmer or gardener. This activity is characterized by a great use of animal manures and sugarcane filter cake mud (*cachaza*), while the processing of urban agricultural wastes to turn them into organic fertilizers is still insufficient.

SEEDS

This sub-program is aimed at regional self-sufficiency of seed production and distribution, which is critical to the success of urban agriculture and without which production would not be stable or sustainable, because it is essential to have the right seed at the appropriate time for sowing. A network of provincial seed farms has been created, whose job it is to keep the supply of seeds flowing. For seeds that are easily produced by farmers and gardeners, such as cucumber and cowpea, the goal of this program is to make production units self-sufficient for such crops. This has now been achieved in all urban farms.

ANIMAL FEED

The jump in small-scale animal rearing cannot be maintained by solely recycling agricultural waste products as feed. To maximize production of animal protein per unit area, a program was created to supplement the use of all household scraps and crop residues with the production of feeds on urban farms. Typically these feeds consist of grains, tubers, roots, and sugarcane. Despite some progress, most units are still not self-sufficient in terms of animal feed.

SCIENCE, TECHNOLOGY, AND TRAINING

Training of urban farmers is critical to perfecting the production technologies being employed. In the training sub-program we focus on practical training, which takes place right in the garden plot, raised beds, or animal rearing pens. We have built an extension system, which counts on the participation of its own extension agents, plus research centers, the most experienced farmers and gardeners, and other individuals and institutions related to urban agriculture. Extension is at all times tailored to local conditions and needs, providing farmers with the latest theoretical and practical information.

OTHER SUB-PROGRAMS

The remaining programs are all of recent origin. These include the sub-programs on sheep and goats, bees, aquaculture, small-scale "popular" plantain production, trees, coffee and cocoa; tropical roots and tubers; oil seeds; irrigation; small-scale agro-industry; and land use. In most regions these programs are still in their formative stages, some more advanced than others (e.g., rabbit rearing in the western provinces of the country).

Key Issues in the Development of Urban Agriculture in Cuba

As urban agriculture has grown in Cuba, it has become apparent that there are several key factors that must be borne in mind.

CONSERVATION AND MANAGEMENT OF SOIL FERTILITY: The productive potential of land available for food production is directly correlated with soil fertility. Although there are many factors that are important in fertility conservation, some require greater attention than others under actual field conditions. Central among these is the control of erosion—maintaining the structure and the physical condition of the soil. The intensity of rainfall in Cuba leads to the rapid leaching of soil nutrients and organic matter, and causes physical damage to soil structure and planting beds. It has proven essential to use a variety of agronomic techniques to protect soils from the effects of erosion. The periodic application of organic matter to soils, planting beds, and containers is also indispensable, as nutrients lost or removed by the previous harvest must be returned or recycled, building the fertility necessary for future sowings (Peña 1995). Finally, appropriate crop rotations and pest management systems adapted to local conditions have been essential.

INTEGRATED PEST AND DISEASE MANAGEMENT: Pest and disease management is based mainly on cultural techniques and biological pesticides. The former rely principally on site selection and planting dates, crop varieties resistant to pests and diseases, adequate soil management, the elimination of alternate hosts of pests and diseases, crop rotations, elimination of infected plants, and thinning and pruning. During the spring and summer months when temperatures are at their hottest, seedling production is carried out in shadehouses, and the technique is employed of leaving clods of earth on the roots of the seedlings to be transplanted. INIFAT has developed totally organic seedling technology guaranteeing high-quality transplants with high yield potential, based on the local resources available in each area. This technique has reduced pest and disease problems due to the high level of vigor displayed by the transplants. The use of biopesticides and other biological pest controls is still being perfected, both in terms of guaranteeing an adequate and opportune supply, as well as with regard to application techniques. *Bacillus thuringiensis* and *Beauveria bassiana* have entered common use since the development and spread of artisanal production at the Centers for the Production of Entomophages and Entomopathogens (CREEs), and they are used against a variety of pests. *Trichoderma* spp. is used for the control of soil diseases. The introduction of new technologies such as neem (*Azadirachta indica*) extracts, and their artisanal and semi-industrial production, as well as of new biofungicides with demonstrated effectiveness, are critical for urban agriculture. Overall, food production in the cities is characterized by low pest and disease incidence, thanks especially to small plot sizes and the generous application of organic matter to the soil.

CROP–LIVESTOCK INTEGRATION: The nature of food production in cities pushes us towards high levels of production per unit area, facilitated by high levels of agrobiodiversity. The highest levels of productivity in organic farming are obtained when crop and livestock production are linked and fully integrated, a task which compels researchers, farmers, and extensionists to work in the closest possible degree of collaboration. Already over half of the urban farms have effective linkages between crop and livestock production.

Urban Agriculture and Sustainability

The organic farming practices used in urban agriculture do not in and of themselves guarantee sustainability. To achieve sustainability all aspects of production must be rationalized and integrated, so that each component complements each other component, in such a way that each action leads to a better outcome at a lower cost.

The best example of this is found in the use of harvest residues and unmarketable portions as animal feed, in turn using the animals' manure to fertilize the crops. We have developed a set of indicators of sustainability to use in perfecting urban production systems. Among these indicators are the following:

- amount of organic matter collected, processed, and applied
- use of soil conservation methods to prevent erosion
- degree to which seeds and starter-animals (i.e. chicks) are produced locally
- degree to which varieties and breeds are adapted to local conditions
- degree of crop–livestock integration
- local water availability and soil moisture
- efficiency of water use
- amount of food produced/hectare per year
- amount of food produced per capita
- use of integrated pest and disease management systems
- net profitability of production
- degree of participation of farmers in training courses and extension activities

By keeping track of these indicators we can monitor the development of urban agriculture. As the indicators improve over time, the sustainability of urban agriculture will be consolidated.

Conclusion

Over past few years the urban agriculture movement in Cuba has clearly demonstrated the food producing potential of cities. Today it is an important source of food for our urban populations. This has been made possible by the decisive effort put forth by urban farmers, and by the support given them by the government to carry out their tasks. The high-level of organization that has been achieved should make it possible to successfully implement the ambitious plans that are currently on the drawing board. We expect that in the near future urban agriculture will satisfy a high percentage of the food needs of our population.

Notes

1. Editor's note to the English edition: At the time of the final editing of this volume, an estimated 90 percent of the fresh produce consumed in Havana is being produced in and around the city (Egidio Páez, personal communication).

2. One quintal equals 100 pounds.

References

Carrión, M. 1996. Agricultura del Hogar en La Agricultura Urbana y el Desarrollo Rural Sostenible. MINAG-FIDA-CIARA: 58–72.

Companioni, N., A. Rodríguez Nodals, E. Fuster, M. Carrión, E. Peña, R.M. Alonso, M. García, and A. Martínez. 1996a. *La Agricultura Urbana en Cuba. La Agricultura Urbana y el Desarrollo Rural Sostenible.* MINAG-FIDA-CIARA: 9–15.

Companioni, N., M. Carrión, E. Peña, and Y. Ojeda. 1996b. Los Fertilizantes Orgánicos: Vínculo fundamental entre la crianza de animales y los cultivos. *Agricultura Urbana.* Primera Reunión Regional sobre Disminución del Impacto Ambiental de la Producción Animal Intensiva en Zonas Peri-Urbanas. Dominican Republic: FAO-JAD.

Companioni, N., A. Rodríguez Nodals, M. Carrión, R.M. Alonso, Y. Ojeda, and E. Peña. 1997. La Agricultura Urbana en Cuba. Su participación en la seguridad alimentaria. *Conferencias.* III Encuentro Nacional de Agricultura Orgánica. Villa Clara, Cuba: Central University of Las Villas: 9:13.

Figueroa, V. and J. Lama. 1997. *Manual para la Conservación de Alimentos en el Hogar.* Proyecto Comunitario Conservación de Alimentos. Havana.

Figueroa, V. and J. Lama. 1998. *Cómo Conservar Alimentos y Condimentos con Métodos Sencillos y Naturales.* Proyecto Comunitario Conservación de Alimentos. Havana.

Figueroa, V. and J. Lama. 1999. *El Cultivo de las Plantas Condimentosas y su empleo en la Cocina.* Proyecto Comunitario Conservación de Alimentos. Havana.

GNAU. 2000. *Lineamientos para los subprogramas de la Agricultura Urbana.* Havana: Grupo Nacional de Agricultura Orgánica, MINAG.

López, F. 2000. El país espera por la respuesta de los orientales en el año 2000. *Granma,* January 26:2.

MINAG. 1994–1999. Informes Anuales 1994, 1995, 1996, 1997, 1998, 1999. Comisión Nacional de Organopónicos y Huertos Intensivos. Havana: Grupo Nacional de Agricultura Orgánica, MINAG.

MINAG. 2000. *Manual Técnico de Organopónicos y Huertos Intensivos.* Havana: INIFAT, GNAU.

Ojeda, Y. 1997. Impacto Económico Social del Extensionismo Agropecuario en la Agricultura Urbana. XI Fórum de Ciencia y Técnica, INIFAT. Havana.

Ojeda, Y. 1999. La Granja Urbana: Elemento facilitador del desarrollo de la agricultura urbana. I Fórum Tecnológico Especial de Agricultura Urbana. Nivel Provincial.

Peña, E. 1995. Cachaza como Sustrato en Organopónicos. II Encuentro Nacional de Agricultura Orgánica. Havana: ICA.

Peña, E. 1998. Uso de diferentes dosis de materia orgánica en los cultivos de lechuga y tomate En VII Jornada Científica La Agricultura Urbana en Cuba. Estructura y Fundamentos Orgánicos. Havana.

CHAPTER FOURTEEN

"Cultivo Popular": Small-Scale Rice Production

Miguel Socorro, Luis Alemán, and Salvador Sánchez, Rice Research Institute

The "Special Period" brought unprecedented difficulties to the large-scale flooded systems that had dominated Cuban rice production for years. In response to declining production, and hence scarcity, small-scale farming of dryland rice, or what is called *cultivo popular* (loosely translated, "popular" or "people's" rice), has sprung up across the country and is increasingly important in national production.

Popular rice production differs from conventional growing in that it relies mainly on local resources and utilizes little to no chemical fertilizers, herbicides, or pesticides. These substances have been very scarce since the early 1990s.

Rice Production in Cuban Agriculture

Rice is one of the most basic components of the Cuban diet. It is usually white and refined, and it is often cooked together with either red or black beans into a popular Cuban dish called *congrí.* Official statistics on commercial rice sales, total inhabitants, and noncommercial consumption (in nursery schools, schools, nursing homes, hospitals, etc.), indicate that annual rice consumption in Cuba exceeds 44 kg annually per capita (Castillo 1997). This means that daily consumption is approximately 120 grams, supplying the nutrients described in Table 1 as well as protein, vitamin B1, and niacin. Consumption is thought to be even higher among the rural, farming population who traditionally cultivate their own rice for consumption within the family unit. Farming families probably consume two times more rice than the official estimates. Given the importance of rice as a source of calories and nutrients, rice production is a fundamental component of Cuban agriculture.

Table 1: Nutritional content of estimated daily per capita rice consumption (120 g)

NUTRIENT	AMOUNT PRESENT	% DAILY RECOMMENDED ALLOWANCE
Calories	433 k cal	18%
Protein	7.8 g	10.8%
Fat	1.2 g	1.6%
Vitamin B1	0.096 mg	8.0%
Vitamin B2	0.024 mg	1.6%
Niacin	1.8 mg	10.6 %
Folic Acid	12 mg	5.3%
Iron	0.6 mg	4.3%
Calcium	4.8 mg	0.6%

(Juliano 1985)

Cuba has always striven to be self-sufficient in its domestic rice production. However, even during the peak years of agricultural production during the mid-1980s, it was never able to produce enough to cover more that 60 percent of demand and has always relied on imports from abroad. Since the onset of the economic crisis in 1989, large-scale rice production has fallen dramatically due to decreased area under cultivation and lower harvests per hectare, with production falling to account for only 20–25 percent of consumption, increasing the reliance on imports to satisfy demand.

The History of the Popular Rice Program

Popular rice production began during the 1980s when the farmers' markets were closed down and rice was largely unavailable through the parallel market. People began to plant rice themselves, generally in small plots with poor soil quality and limited access to water. This grassroots production increased after 1989 when the economic crisis began.

At the beginning of the Special Period, the Cuban government initiated the Popular Rice Program in order to support grassroots cultivation. In 1992 and 1993 popular production increased, particularly in Cuba's western and central provinces. When the farmers' markets reopened in 1993, popular cultivation rose as the soaring prices for rice drove more individuals to produce for sale or subsistence. Individuals and families often grew rice to satisfy their own dietary needs; however those with more land were able to market their surplus for a healthy profit.

The Popular Rice Program also includes rice production that is conducted in agricultural enterprises whose primary crops are not rice. Many large farms began to produce rice on smaller plots in order to supplement their standard production. This rice could be used in their cafeterias and sold to their workers, and any surplus was sold in the farmers' markets.

Most new rice farmers who seriously applied themselves to this crop achieved successful harvests even if they had little experience. Their production has now had a cumulative effect on the availability of rice to the general population, though the price still fluctuates with supply, depending on production and harvest cycles. However, as production has increased, the price of rice in the now reopened farmers' markets has remained between two and five Cuban pesos per pound (much lower than before).

Organizational Structure of the Popular Rice Program

The Popular Rice Program is structured in a multi-tiered system that allows for national dissemination of information, technologies, and research while giving local governments control over the particular production decisions related to their regions. Since inception, the program has been under the aegis of the National Association of Small Farmers (ANAP) and the rural sector of the Ministry of Agriculture (MINAG). Together they began to organize a multitude of individual initiatives and to implement a cohesive, national program.

MINAG has worked to increase popular production by enacting a series of land reforms. It has made small plots of land available in usufruct (the land

Technician from the Rice Research Institute and campesino *share experiences in evaluating new varieties.*

is leased for free on condition that it remains under production) to those who wish to begin popular production. The large number of applicants include urbanites, suburbanites, agricultural workers, and retirees, many of whom have been able to plant and harvest their own rice. These individual initiatives have played an essential role in the recovery of rice production while at times providing local employment and generating economic growth.

However, the Popular Rice Program is, by definition, based on grassroots initiatives that originate outside of the standard agricultural structure. Responsibility for the program has been localized in the hands of the smallest municipal governing body, called the Popular Council. The councils identify possible areas for cultivation and balance priorities, always moving towards the final goal of community self-sufficiency. The Popular Councils, because they are completely integrated with the community, are better able to make decisions regarding the balance between individual and social needs and interests.

The results of the program have varied from province to province, mainly because of geographical and climactic differences that dictate the type and extent of production. Provinces such as Pinar del Río, which still contain large-scale and highly productive rice farms, plant 15,000–16,000 hectares of rice per year, and produce more than 26,000 tons for domestic consumption. Provinces such as Ciego de Ávila have smaller farms and tend to plant 6,000 hectares of rice and produce approximately 6,000 tons. The province of Santiago de Cuba, and other highly urbanized areas, rely primarily on popular rice production and tend to plant only 600–800 hectares of rice per year with harvests averaging around 700 tons.

Of these totals, an estimated 90,000–100,000 hectares nationwide are dedicated specifically to popular rice production, generating 110,000–130,000 tons per year. This estimate is not exact, given the difficulty of quantifying the individual production of each province.

General Technological Characteristics

Until the 1990s, almost all of Cuba's rice production was large-scale and done in lowland areas. The fields were heavily irrigated so as to grow the rice under a layer of water, and cultivation required specialized technology and expensive inputs. However, popular rice farmers do not always have access to these expensive resources. Their cultivation is on a much smaller scale, rarely mechanized, and more dependent on the existing ecology because it uses fewer inputs and is done on land not traditionally considered suitable for rice. Because Cuba can be subdivided into a vast number of microclimates—geography, temperature, rainfall, and soil quality vary from region to region—

it is essential for popular production that agricultural technologies are adapted to be appropriate to the specific regional contexts (Socorro et al. 1997a).

To respond to the specific needs of this new, low-tech, small-scale production style, in 1996 MINAG commissioned the Rice Research Institute (IIA) to work with producers to develop appropriate technologies. The goal was to further stimulate small-scale rice production by increasing the amount of land under cultivation and the per-acre output of existing plots without damaging the environment or causing financial hardship for producers.

An important concern of the IIA is that, since the agricultural transition, more rice is being grown in drier conditions that depend primarily on rainfall for irrigation. Today:

20 percent of rice plots are irrigated during the dry months (December–April)

40 percent of rice plots are irrigated during the rainy months (May–August)

40 percent of rice plots are not irrigated (rely on rainfall during May–August)

Approximately 40 percent of popular rice cultivation falls under the final category and uses no irrigation other than natural rainfall. The practice of dry rice cultivation has affected the choice of technologies and seed varieties. Like Byerle in 1994, a 1998 study by Alfonso found that the IR-1529, Perla, IA Cuba-23, and IA Cuba-24 varieties performed better under dry conditions.

Popular rice cultivation has also proven that rice can be grown in mountainous areas, and during the off-season. In the eastern and central provinces of Cuba, farmers have been able to plant rice in September and October because of the temperatures and rainfall patterns that are unique to these areas. This has been particularly fruitful in the Baracoa region of Guantánamo province, at the eastern tip of Cuba (Aira 1997).

Western Cuba has contributed to new techniques for popular rice cultivation by introducing the practice of transplanting seedlings into the field. This technique is widely used in Asia and has become popular in the provinces of Pinar del Río, Havana, Matanzas, Cienfuegos, and Villa Clara. This differs from the direct sowing technique used in the United States, Australia, Italy, and other countries because it practically eliminates the need for herbicides. The fields are covered with a layer of water, thereby killing any existing weeds, and the seedlings are not planted until they are at least 25 days old and large enough to eliminate competition from weeds as long as the layer of water is maintained. With direct sowing of seed, two to three different herbicides would be needed to control weeds, thereby contaminating the environment with toxic chemicals. Transplanting also guarantees a more uniform crop that boosts yields to between five and ten tons per hectare (Lerch 1972). Conventional rice cultivation can yield up to ten tons per hectare, but only with extensive application of expensive chemical inputs.

Farmers have also been able to increase yields by moving away from mechanization. Depending on seed variety, region, and time of year, some farmers who harvest by hand are able to obtain a second harvest from the sprouts left behind after the first harvest. The second harvest can produce 30–40 percent of the total produced by the first harvest in a very short period of time.

Another potential for increasing the effectiveness of popular rice cultivation is the use of sesbania (*Sesbania rostrata*) as a green manure. The Rice Research Institute has been studying and introducing this legume, which thrives in conditions similar to those of rice plants. Sesbania has a demonstrated ability to fix atmospheric nitrogen, similar to the bacteria *Azorhizobium caulinodans.* It produces a large number of nodules in the stem and branches that can supply as much as 80 kg of nitrogen per hectare for the subsequent rice harvest. Sesbania grows very rapidly in high temperatures, producing 30–40 tons/hectare of biomass in only a few months. Its rapid and deep root growth enhances soil structure and adds phosphorus and potassium to the soil. This plant has great potential to act as a fertilizer for rice cultivation without producing any environmental contamination or requiring expensive inputs or technology.

Cuba is currently exploring other alternative technologies in order to make popular rice cultivation more sustainable. One such technology is the use of animal traction to prepare soil for planting in both irrigated and dry conditions. The first steps have been taken to introduce buffalo-driven plows similar to those of Vietnam, China, and other Southeast Asian countries.

After harvest, most of the rice that is grown through the Popular Rice Program is processed using low-tech, artisanal techniques. Rice is cleaned using the wind, and then is sun-dried, and husked and polished in small, privately owned mills. Current rice production generates thousands of tons of byproducts such as bran, which is used as feed for poultry and swine, thereby contributing to the production of animal protein for human consumption.

Rice Seed Varieties

The use of appropriate seed varieties is considered fundamental to the sustainable development of popular rice agriculture. Cuba is currently employing an assortment of rice seeds and is seeking to expand the program by donating seeds to farmers. Each farmer would receive the seed varieties most appropriate to the conditions of his or her farm. Seed varieties would vary depending on the amount of available irrigation and salinization, require

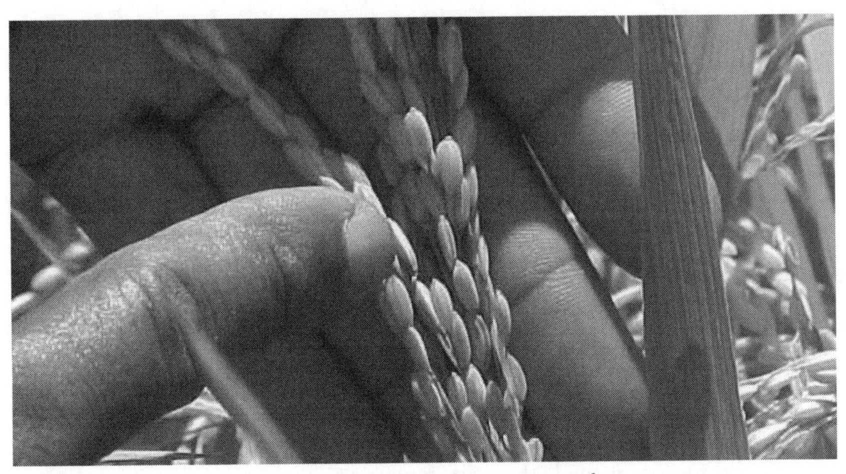

Rice variety with promising characteristics for low-input conditions.

little artificial fertilization, and be highly resistant to the principal pests and diseases found in each region.

The wide variety of seeds used in Cuban rice cultivation is primarily due to the work of the Rice Agro-Industrial Complexes (CAI), state-run businesses that specialize in rice production. The CAIs supply farmers in their region with seeds. The most popular varieties are Amistad-82, Perla de Cuba, Caribe-1, and J-104, but in the more remote areas farmers still cultivate more traditional varieties, sown in Cuba for almost forty years, known only by their local names. These varieties are similar to Honduras, *Patipreito*, and Blue Bonnet varieties and are characterized by their height of over a meter and the fact that, while their yields are low, the rice is of very high quality.

The Rice Research Institute is currently testing a number of high-yield varieties in a wide range of conditions in an effort to increase production per hectare. Producers have already found that certain native varieties such as IACuba-18, IACuba-20, IACuba-25 LP-2, LP-6, and LP-7 produce and function well in the fields.

The Future of Popular Rice

The ongoing mission of the Popular Rice Program is to continue to increase domestic rice production. This can be done by increasing yields per hectare and by placing more land under cultivation. A significant amount of exploitable land still lies fallow, especially in the provinces of Las Tunas, Holguín, and Guantánamo, where the program has not been as fully developed. These eastern provinces are mountainous and have less available water for agricultural use, therefore most of the rice will be cultivated without irrigation. The rice research centers are focusing on developing more rice varieties appropriate to the conditions of eastern Cuba.

In order to enable more farmers to cultivate rice, the Popular Rice Program engages in training and extension and the diffusion of appropriate technologies. Many provinces contain Rice Research Institute training programs that have trained over 300 producers and twelve agronomists in 1997 and 1998 (Socorro 1997b). The agronomists are now responsible for conducting further educational seminars in other areas of the country. The principal content of the training courses is:

- how to select land for rice cultivation and plan and construct a system of irrigation and drainage
- methods of transplanting seedlings which save seeds by using up to five times fewer seeds than direct sowing
- when to plant in order to maximize harvests
- which varieties are best suited for local conditions
- rational use of water and herbicide reduction
- alternative sources of nutrients that do not require chemical fertilizers
- when and how to harvest the rice in order to minimize losses while gathering and utilizing the byproducts of rice

Conclusion

With further training and expansion, the Popular Rice Program has the potential to equal or exceed in size the 150,000 hectares under conventional cultivation. However, given the financial limitations on popular production, the full potential of popular rice has not yet been realized. Cuba must continue to explore new technologies and develop ways to overcome these obstacles, ensuring an increase in the supply of this staple food. A realistic production goal is to produce 35 percent of the domestic demand for rice on Cuban soil while maintaining the commitment to sustainability and low production costs.

References

Aira, A. 1997. Resultados obtenidos en el cultivo popular de arroz en la provincia de Guantánamo durante 1997. V Reunión Nacional de Popularización del Arroz. Morón, Cuba.

Alfonso, R. 1998. Determinación de parámetros genético-fisiológicos indicadores del estrés hídrico para su empleo en el mejoramiento genético del arroz (dissertation). Havana: ISCAH.

Byerle, D. 1994. *Modern Varieties: Productivity and Sustainability, Recent Experiences and Emerging Challenges.* Mexico City: CIMMYT, pg. 26.

Castillo, D. 1997. La calidad del arroz en Cuba. Ponencia X Reunión Internacional para arroz en América Latina y el Caribe. Acarigua, Venezuela.

Juliano, B.O. 1985. Polysacharids, proteins and lipids of rice. *Rice: Chemistry and Technology* (second edition). American Association of Cereal Chemists, pp. 59–174.

Lerch, G. 1972. Desarrollo y rendimiento del arroz variedad IR-8 en Cuba. *Serie Agrícola.* No. 21. Havana: ACC, pg. 96.

Socorro, M., L. Alemán, S. Sánchez, G. García, C. Pelaez, and A. Aira. 1997a. La investigación-desarrollo en el contexto del Programa de Popularización del Arroz. Ponencia Coloquio Franco-Cubano: Métodos y Experiencias en Extensión Agropecuaria para un Desarrollo Sostenible. Holguin, Cuba, pg. 5.

Socorro, M., S. Sánchez, F. Cruz, A. Hernández, and R. Sanzo. 1997b. Desarrollo de la comunicación en el Programa de Popularización del Arroz. Ponencia Coloquio Franco-Cubano: Métodos y Experiencias en Extensión Agropecuaria para un Desarrollo Sostenible. Holguin, Cuba, pg. 4.

CHAPTER FIFTEEN

Sugarcane and Sustainability in Cuba

Rafael Suárez Rivacoba, Ministry of Sugar, and Rafael B. Morín, National Standards Office

For centuries, sugar has been one of the most important and universally consumed components of the human diet. This is both because it is a low-cost energy source, and because of its potency as a sweetener. Worldwide, sugar can be produced under the most diverse climatic conditions. In temperate regions, sugar is extracted from beets, and in tropical and subtropical regions from sugarcane.

In recent years, annual world sugar production has been approximately 120 million metric tons, with 60 to 65 percent derived from sugarcane, and the remainder from beets. These figures do not include uncentrifuged sugar production, which has economic and social importance in countries such as Colombia and Ecuador, as well as in India and other Asian countries.

For many years sugar producers have faced a critical situation in international trade, due to a relative closing off of the so-called "free market," and to prevailing world market prices that are below the cost of production. The causes of this situation are clear and widely known, and arise from the protectionism of developed countries, mainly the United States and the European Economic Community. The United States has reduced imports from four to five million tons to two million tons annually. The European Economic Community went from being a net importer, to one of the main exporters to the "free market."

Sugarcane's genetic yield potential has scarcely been tapped, and it can be cultivated by techniques that are more appropriate and sustainable than those which are common in today's world, in both economic and ecological terms. Improvements can be made to techniques that until now have been "imported" from developed countries, and are based on intensive chemical fertilizer and pesticide use.

Sugarcane is a plant with exceptional characteristics, capable of synthesizing soluble carbohydrates and fibrous material at a higher rate than other commercial crops. This ability opens up practically an infinite number of uses for its hundreds of by-products, which in many cases are of greater added value and economic importance than sugar itself.

Agricultural residues and industrial effluents from sugar production, far from being environmental pollution factors, can play an important economic role, as well as being used to restore vital elements to the soil.

Sugarcane Agriculture

Currently, commercially cultivated sugarcane varieties are species and hybrids of the *Saccharum* genus of the cereal family (*Poaceae*). Sugarcane is what is known as a "permanent" crop, harvested at periods of between 12 and 24 months. An average planting may last from 5 to 10 cuttings, although this varies between regions, and depends on agricultural practices.

Sugarcane and sugar production form integral parts of the history, culture, and traditions of the Cuban nation. Cuban history has been characterized by social injustice, with its cruelest expression experienced during the colonial period when African slaves were brought to work on sugarcane plantations and in sugar mills (Guerra 1961; Moreno-Fraginalis 1978).

Since 1959, the modernization process in sugarcane agriculture has been directed towards a more intensive agriculture system with high inputs, mainly

Cuban harvester cutting sugarcane.

characterized by the introduction of mechanization. By the end of the 1980s, mechanization covered 75 percent of sugar cutting and 100 percent of collection, while average national production was about 7.5 million tons per year.

The beginning of the 1990s brought about the collapse of European Socialist Bloc. Cuba was faced with extremely difficult circumstances. For the first time in three decades, sugar production, of which 90 percent was exported, was completely in the hands of the "free market," in which products are sold at extremely low prices due to the protectionist measures of most developed countries.

The lack of financing capacity or credit to purchase necessary inputs brought about the start of a transformation of farm technology in Cuba. We had to find or develop alternatives that make maximum use of the ecological characteristics and potential of each crop, including its interactions with the environment.

Sugarcane has extraordinary capacities compared to other plant species. Under good conditions it can produce 100 tons/ha of stem biomass, plus about 20 percent more consisting of leaves and growing tips. Sugarcane is so productive in part because it has much more efficient photosynthesis than other crops, allowing it to make better use of solar radiation, and equally, to absorb more carbon dioxide from the atmosphere (Alexander 1985, 1986).

For that reason alone, sugarcane is an important ecological factor on a global scale, which contributes to reduction of environmental heat produced by the "greenhouse effect." Although there are no definite studies available on the photosynthetic activity of sugarcane, it has been estimated from basic studies that it has a carbon fixing capacity greater than two tons per hectare per year (Rhodes and Rank 1991). This is comparable only to the carbon fixing capacity of deciduous forests in temperate regions (Acosta 1992).

Since the beginning of the 1990s the aim in Cuba has been to attain a more natural or organic sugarcane product with a lower dependence on expensive chemical inputs. Such production would be based on the most modern scientific and technical developments, and display true ecological and economic sustainability.

The harvesting of green cane has been gradually introduced in Cuba, and now covers approximately 90 percent of our sugarcane regions. Pneumatic machines are used that partially clean the sugarcane. The straw is left on the soil as a protective mulch to preserve moisture, reduce erosion, and contribute to weed control. This alone has reduced the use of herbicides by 35–50 percent, and decreased tractor use for soil preparation by up to 33 percent (Díaz-Casas 1996). This has had a tremendous impact, both technically and economically: while protecting the environment by reducing fuel used for soil preparation, it has also reduced the use of toxic chemicals and brought down production costs. A beneficial ecological effect also results in a lessening of air pollution and soil degradation when the cane is not burned.

Sugarcane is cleaned for the second time at collection centers, where it is then transferred to railway cars that take it to the sugar mills. A significant part of the residues from cleaning are used as fodder for cattle, of great importance since this feed is produced during the dry season when there is very little fresh pasture for grazing.

Another portion of the residue is used as fuel in the *bagasse* boilers, and some is used as a raw material in compost production. Compost from sugarcane *cachaza* (filter cake mud) is highly valued for its organic matter, and is excellent for improving soil conditions in cane fields. An extensive soil improvement program utilizes microbiological inoculations and cattle manure from herds stabled near the sugarcane centers. The combined application of *bagasse* from the sugar mill, and the use of agroindustrial effluents for irrigation, add important alternative fertilizers. This in turn creates a closed cycle between the sugarcane field, the processing industry, and livestock production.

Because of the lack of fuel and machinery for agriculture, there has been an increase in draft animal use as a substitute. Presently, a series of tasks are carried out by oxen in sugarcane agriculture. In 1997 this reached almost 50 percent of the sugarcane area.

Other alternative agricultural techniques used extensively in Cuba are the minimum tillage of soils (Bouza et al. 1996), crop rotation, and intercropping (Crespo 1996). With minimum tillage using equipment that does not invert the topsoil, not only are soil preparation costs reduced, but there is less soil erosion, and the effect of the intense solar radiation of the tropics is lessened.

Intercropping leguminous peanuts with sugarcane.

Legumes such as beans and soybeans are used in rotations and in intercropping, boosting farm profits and improving soil fertility through nitrogen fixation and other physical and microbiological effects.

The optimum use of fertilizers has been achieved through the Fertilizers and Amendments Extension Service (SERFE) program (INCA 1996). The idea of SERFE is to eliminate unnecessary overuse of fertilizers, applying only the fertilizer that is really needed. This system eliminates the previously common irrational use of fertilizers that affected the environment and increased production costs. In addition, it is designed to maintain the biological equilibrium of the soil. At the same time, fertilization research is being carried out on the use of nitrogen fixing bacteria. Different strains of *Azotobacter, Acetobacter, Azospirillum,* and other microorganisms are being actively tested in production.

Another important method for soil conservation is contour planting, which fights erosion and is being adopted in sugarcane fields with slopes of greater than eight percent (Cuba has 228 thousand hectares with that characteristic) (Alvarez 1995). When this technique is employed, yields have increased to an average of 13 tons per hectare. The national program is working toward applying this technique in all areas that have these physical requirements.

Pest and disease control is carried out largely by using resistant varieties. Biological control is employed with the release of beneficial insects and microorganisms. The Cuban varietal selection program guarantees that new strains are substituted for varieties that are degenerating and beginning to be susceptible to pests and diseases. Presently, 92 percent of the sugarcane varieties in production are of Cuban origin, with only 8 percent coming from outside the country. The imported varieties are submitted to strict field studies, analyzing performance and adaptation to the soil and climate types of Cuba. They are accepted only if yield and other indicators are equal to or surpass the Cuban varieties.

The program has a germplasm bank with more than 2,500 accessions, 14 regional stations, 63 experimental blocks in commercial areas, a system of certified seeds, with a seed bank in each province, and a quarantine station. Accelerated micro-propagation is used to complement traditional seed reproduction methods. Presently, approximately 10 million in-vitro seedlings are produced annually in our bioplants.

Of first priority in biological control is the fight against the sugarcane borer (*Diatraea saccharalis*), which produces the greatest losses in Cuba. Fifty Centers for Production of Entomophages and Entomopathogens (CREE) annually release millions of *Lixophaga diatraea* fly pupae in sugarcane plantations. Other natural enemy strategies are being tested, such as using *Aphanteles* and *Trichogramma* wasps, sterile male releases, and microbial biopesticides.

Organic Sugar

The metamorphosis of the sugarcane industry has made Cuba's conversion from traditional to organic techniques possible, and in a relatively short time. Nevertheless, beginning on a small scale was required. Commencing in 1997 near the Central University of Las Villas (UCLV), organic cane production without chemical inputs was implemented. Sugarcane was processed in a pilot plant owned by the university, and now up to one thousand tons are produced per year for export to the European market (which received their first significant quantity of Cuban organic sugar in 2000). ECOCERT International, a German organization, will certify this sugar.

The initiation of this environmentally friendly technology is the result of a coordinated effort between the Organic Farming Group (GAO-ACTAF), the Ministry of Sugar (MINAZ), and the Central University of Las Villas. Special support for this work has been supplied by the Italian Association for Organic Agriculture (AIAB). AIAB trains and teaches specialists who work in Cuba, and complies with the standards and rules (2092/91) established by the European Economic Community for organic products.

The successful development on a small scale by the Central University has given sufficient necessary background information to the Ministry of Sugar to enable the beginning of industrial-scale production. To achieve this, an organic sugar Agro-Industrial Complex (CAI) will be created, having an average processing capacity of 4,600 to 6,900 tons per day. This sugar mill will process only organic sugar in order to produce a high-quality product for export, and to provide a better product to Cuban consumers.

Some of the by-products from this production will be recycled as organic manure in agricultural areas, and in other cases might be used for producing processed organic products in popular demand. These technologies and applications, supported scientifically, allow for a progressive approach towards effective agricultural and industrial integration, and make for a truly sustainable sugar crop.

Diversification of Uses for Sugarcane

More than 400 years of use as a raw material for producing sugar has created a mentality that sugarcane has no other uses. Yet about two-thirds of the 250 million tons of sugarcane produced annually in the world are used to produce ethyl alcohol for automobile fuels and additives.

Diversification means the integrated use of the sugarcane plant, the use of sugar production by-products, and the use of sugar and sugarcane as raw materials for transformation into other economically and socially important products. It also refers to optimum land use, for example, through intercropping or crop rotation. In Cuba the diversification of sugar does not mean its replacement with other crops. Thus, diversification is not just a strategy to confront the sugar market crisis, but a pathway to industrialization and social development in Third World countries.

To make integrated use of the biomass produced by the sugarcane plant, one must go beyond the stems—the portion used for sugar production—and consider the leaves and tops. In most of the world sugarcane tops and leaves (fresh or dry) are burnt or discarded at harvest time. Table 1 shows the chemical composition of both portions.

Table 1. Composition of sugarcane (%)

COMPONENTS	STALKS	TOPS AND LEAVES
Dry matter	29.0	26.0
Total sugars	15.4	0.2
Sucrose	14.1	0
Lignin cellulose	12.2	19.8
Ash (fiber)	0.5	2.3
Other components	0.8	2.4
Water	71.0	74.0

(Noa 1982)

The ratio of stalks to tops and leaves depends on the variety planted, the agronomic practices employed, and the age of the sugarcane at harvest. Studies carried out in Cuba demonstrate a direct relationship between this ratio and agricultural yield, as shown in Table 2.

Table 2. Yield components of the sugarcane plant

YIELD (T/HA)	STALKS (%)	OTHERS (%)
30–50	78	22
51–70	79	21
71–85	80	20
More than 85	82	18

(Casanova 1982)

With current sugar processing techniques, every 100 tons of clean stalks going to a sugar mill yields a variety of by-products, as shown in Table 3. Thus there is a question of making adequate use of the sugarcane and of its by-products. Soluble sugars, mainly sucrose, glucose, and fructose, can be extracted in aqueous solutions in mills or diffusers, as is done in sugar or alcohol production. The chemical or biotechnological transformation of these sugars into other products could fill an endless list. Table 4 shows some of the most important commercially produced products.

Table 3. By-products obtained from processing 100 tons of clean sugarcane stalks

PRODUCTS	QUANTITY (TONS)
Sugar (96% Pol)	11.0
Bagasse (50% moisture)	27.5
Final molasses (88% solids)	3.5
Cachaza (75% moisture)	2.5
Other residues	20.0

(Casanova 1982)

Table 4. Some commercial products from sugarcane in Cuba

PRODUCT	PROCESS		USE
	CHEMICAL	BIOLOGICAL	
Sugar	X		Food industry
Glucose	X		Food and pharmaceutical industry
Fructose	X		Food and pharmaceutical industry
Ethyl alcohol		X	Fuel, drinks, chemical industry
Yeasts		X	Feeds, bakery, drinks
L-lysine		X	Animal feed and foods
Citric acid		X	Food and pharmaceutical industry
Lactic acid		X	Chemical and pharmaceutical industry
Acetic acid		X	Different industries
Oxalic acid	X		Chemical and construction industry
Monosodium glutamate		X	Food industry
Acetone and butane		X	Chemical industry
Dextran and xanthan		X	Thickeners
Sorbitol and manitol	X		Food and pharmaceutical industry

In general, the profitability of the derivatives strongly depends on the price of intermediate molasses products. Depending on the price structure, final synthesis can be performed using different intermediate products, or even sugar, as the starting point, as is done in Cuba for the production of dextran, glucose, fructose syrup, and biodegradable tensoactive agents (Cordovés 1978).

The lignin cellulose part of sugarcane, both the *bagasse* that comes from the juice extracting process, or the straw (leaves and tops), can be used in diverse applications. *Bagasse* is, above all, a fuel which is used in sugar factories, and can provide all the energy necessary for the running of a plant and still leave a surplus of up to 30 percent to be used for other purposes. Sugarcane straw can also be used as fuel in sugar mills or other industries. In addition, properly processed by chemical, physical, or biotechnological methods, both *bagasse* and straw produce an excellent high-energy feed for cattle.

A great number of products with greater added-value can be obtained from *bagasse;* some are already produced on a commercial scale in different countries. For example, *bagasse* may be substituted for wood for a number of different applications, where it may offer technical and cost advantages. Different sorts of particle board, paper, cardboard, cellulose, xylose, and furfural, can be obtained from *bagasse.* The price of these by-products is generally based on their energetic value and on the prevailing price of the fuel needed to substitute for that amount of *bagasse* in the sugar mill. For example, 5.2 tons of *bagasse* (50 percent moisture) would be equivalent to a ton of fuel oil.

Considering the limited availability of forest resources, *bagasse* is of special economic and social relevance for many sugar-producing countries as a raw material in the pulp, paper, and particle board industries. Such is the case in Cuba, where much attention has been placed on developing these technologies. The potential of *bagasse* as an annual source of renewable fiber will be analyzed later.

Recent scientific and technical developments in the field of biotechnology have opened up a new spectrum of sugarcane by-products. The unsustainable nature of fossil resources, mainly oil, limited forest resources, and the need for forest preservation as well as other factors such as food shortages all contribute to a bright future for sugarcane products.

Third- and fourth-generation by-products are created through the transformation of first- and second-generation by-products into those of higher added value. Sucrose is transformed into raw materials for medicines such as vitamin C by the glucose-sorbitol pathway, and polymers are synthesized, such as polyhydroxy-butyrate and biodegradable tensoactive agents. Cellulose obtained from *bagasse* becomes cellophane, acetates, rayon, carboxy-methyl-cellulose (CMC), microcrystalline cellulose, etc.

Ethyl alcohol obtained by fermentation has given rise to an alcohol-based chemical industry in which plastics (polyethylene, PVC), acetaldehyde, and its derivatives are made. In countries such as Brazil, India, and Peru (GEPLACEA 1988) commercial projects are already showing fairly good results.

The extensive literature reflects an endless number of alternatives; *The By-products Manual* (ICIDCA-UNIDO-GEPLACEA 1985) offers technical and economic data on a great number of products and technologies.

Sugarcane for Animal Feed

Various projections indicate a growing population in Latin America and the Caribbean. In contrast, per capita food production in the region from 1975 to 1986 decreased by approximately eight percent (Preston 1986). These figures highlight the acute state of malnutrition of people in the region.

Sugarcane can function not only as an important source of calories in people's diet, but it could also contribute greatly to livestock production. Many countries already deficient in pastures, cereals, and proteins used for traditional animal feeding methods would benefit from its use. The two main needs in designing a diet for any type of animal are metabolizable energy and protein. The conventional model finds these two ingredients through high-yield grasses, cereals, and other protein products.

Sugarcane and its by-products offer alternative solutions both for metabolizable energy and for protein supply. These alternatives not only have economic importance for tropical countries, but they represent a strategic long-term contribution to alleviate the critical competition between food for people and livestock *viz a viz* cereal and energy consumption (Pimentel 1997; Preston 1997).

The experiences of various countries over the last 15 years have demonstrated an economic advantage to using sugarcane as the main energy source for cattle feeding in beef and milk production (Preston 1977; Murgueitio 1990). These systems are of special relevance for tropical countries during the dry season, the optimum season for the sugarcane harvest, and in turn, the most critical one for pasture and forage availability.

Once some relatively modest requirements of balance are satisfied, diets based on ground whole sugarcane can allow milk production of 10 to 12 liters per day and weight gains above 800 g daily (Alvarez 1986; Preston and Murgueitio 1988). The main requirements are the addition of urea as a nitrogen source, small amounts of true protein (plant, animal, or single-celled organisms) and glucogenic precursors such as those from wheat, rice, and crop residues. Optimum diets will depend on the local availability and prices of ingredients, but experience indicates that feed based on sugarcane are always much more viable than conventional systems given the real conditions of most sugar producing countries.

There is an interesting possibility of increasing the true protein content of sugarcane through the solid fermentation of its soluble carbohydrates. Recent experiences in Cuba have given encouraging results on the use of a product called *Saccharina,* which has a protein content of six to eight percent. This production process uses not only finely ground sugarcane, but also the so-called "tandem *bagasse* pith" which is extracted in sugar factories, and has a fiber and sugar composition very similar to sugarcane itself. The existing infrastructure is used, without affecting the capacity of the factory, and new investments for grinding sugarcane are not necessary.

Molasses and Sugarcane Juice

An alternative for the integrated use of sugarcane, is the separation of the juice containing the sugars from the *bagasse* (Preston 1988). This offers the advantage of more direct use of the specific potential of each substance.

The most widespread use of soluble carbohydrates in animal feeding is in the form of final molasses, which is not only used in sugar producing countries, but also in some developed countries that import it as an animal feed. Research carried out in Cuba since the 1960s has made it possible to develop feeding systems based on substituting molasses for cereals as a metabolizable energy source (Elías 1986).

In the case of ruminants, the successful use of sugarcane and/or molasses as an energy source, depends on adequate supplementation. In practice, the most cost effective alternatives for supplementation use urea as the fermentable nitrogen source, and some protein-containing forage, or combinations of forages and a small supply of protein concentrates (Preston 1977). Thus molasses has been used to feed cattle in Cuba for many years, as well as in other countries (Figueroa and Ly 1990); in molasses-based diets, urea is supplied as a fermentable nitrogen source and the ration is supplemented with forage, minerals, and true protein.

We also have years of experience in research and commercial application of sugarcane molasses as a feed in swine production (Figueroa and Ly 1990). In Cuba, where pork is an important component of traditional meals, and with the climatic restrictions for cereal production, the use of sugar industry by-products takes on special relevance. However, due to its high content of non-sugar substances, feeding pigs with final molasses is a poor use of its energy value (Ly 1989). For the past few years, an intermediate molasses, which is richer in sugar, has been substituted for final molasses. Intermediate molasses is also cheaper, and therefore more cost effective (Pérez 1986).

The direct use of sugarcane juice for pig fattening is also used in various countries. In Cuba, applications have been developed where juice is extracted in small sugar mills near the pig-fattening unit. This variant suffers from low levels of juice extraction (approximately 45 percent), but is possible with simpler facilities, and the sugar-rich *bagasse* can be used for feeding ruminants.

Bagasse and *Bagasse* Pith

A sugarcane factory produces *bagasse* that can be used in different ways. The *bagasse* pith, which is the finest portion obtained from sifting *bagasse*, is a readily available and inexpensive raw material. The use of these products for animal feeding is attractive due to the modest investment required for the construction and operation of the processing facilities. Its main limitation, compared with other agricultural residues, is low digestibility. Two technologies have been used on a commercial scale in order to increase the digestibility of *bagasse* and its by-products. Mixtures of sugarcane *bagasse* pith with final molasses and urea have been used in Cuba for years (Suárez Rivacoba 1987). An alkaline treatment with sodium hydroxide was developed in Cuba in the 1970s, and has been extensively employed. With any of these systems, it is possible to increase digestibility of the materials from 30 to 35 percent to values closer to 60 percent. This makes such materials comparable to other forages and agricultural residues.

A technology has recently been created in which lime (calcium hydroxide) is substituted for sodium hydroxide with excellent results. Cost/benefit studies carried out using these treatments demonstrate that the increase of metabolizable energy obtained compensates for the costs. With the cost of sodium hydroxide and energy constantly rising, the use of lime as an alternative is very attractive (González 1987; González and Sáez 1991).

Just as in the case of sugarcane and molasses, dietary balance is key is making these materials cost-efficient and effective in animal feeds.

Protein from Sugarcane

Compared to cereals, grains, and grasses, the main limitation of sugarcane and its by-products is its near zero protein content. However, the soluble carbohydrates in the molasses or juice can be transformed by single-celled organisms into protein through widely-known fermentation technologies.

At the end of the 1970s eleven factories were established in Cuba, capable of producing from 10 to 12 thousand tons of torula yeast per year from final molasses. Torula yeast is a powdered protein concentrate with a protein content of 45 to 48 percent.

These factories also produce a cream-like product called "protein molasses," which consists of yeast mixed with intermediate molasses in order to obtain a product with a protein content of 15 percent dry weight. An important method of rearing pigs has been developed utilizing this product, with results that are more cost effective than conventional technologies (Figueroa and Ly 1990). The use of the yeast recovered in alcohol factories has also given results for pig fattening.

Crop Residues

Although sugarcane crop residues (leaves and tops) constitute approximately 20 percent of the weight of the plant, their use in most countries is almost nil. In the Cuban sugar harvest system, large quantities of such residues are left in sugarcane collection centers. Presently, 930 stations and collection centers are in operation, producing an average of 40 tons of residues daily, for an overall annual availability of approximately five million tons. A significant part of these residues are used as cattle forage during the dry season. The feed value of these residues, when consumed fresh, is even higher than that of the other lignin/cellulose materials mentioned (*bagasse* and *bagasse* pith). These residues can also be treated to increase digestibility. As was mentioned above, these residues also have an important role in making compost.

Sugarcane and Energy

The relationship between the energetic value of sugarcane biomass, and the energy needed for its growth and harvest, is a 20 to 1 ratio when calculated for high fertilization levels, irrigation, and mechanized harvest. What this means is that, in the worst case scenario (that of high energy-cost production methods), the energy invested in sugarcane production represents *at the most* five percent of the value of energetic output (López and de Armas 1980).

To be practical, of course, this energetic potential has to be expressed in some form of energy used in modern life. Of these, the most important are the generation of electricity, fuel for steam generation in industry, and fuels for automotive vehicles. The traditional sugarcane mill is not a model of energy efficiency; known and tested technologies use only 70 percent of the energy currently consumed in these mills (Correia 1991; Ogden 1990). If the objective of a sugar mill is to produce surplus *bagasse* for by-products, or to sell as fuel to other enterprises, it is important to decrease steam energy consumption and maximize combustion efficiency, but high pressures are not required. Under these conditions lower steam pressures can be used, all the electrical and mechanical energy required could be produced, plus a *bagasse* surplus of 30 percent and more could be generated (de Armas and González 1986).

To reach efficient *bagasse* combustion levels, modern boilers with ovens that use less air are required. Ovens should also be equipped with recovery surfaces for better gas utilization (de Armas and González 1986). Several countries, including Brazil and Cuba, now produce these boilers. One way to improve the efficiency of existing boilers, requiring only a modest investment, is to install *bagasse* dryers that use chimney gases to reduce moisture content from 50 percent (when it leaves the mills) to 25 or 30 percent. In Cuba positive results have been achieved with this system (Arrascaeta 1988).

The new sugar mills built in Cuba in recent years with relatively simple plans and equipment are designed to operate with moderate pressure, and produce 40,000–50,000 tons of *bagasse* per harvest, and seven to eight thousand megawatt hours of surplus electricity. Many sugar mills have been added as suppliers in the national energy grid, delivering energy surpluses that contribute to the reduction of oil imports. Other successful alternatives have included the use of harvest residues as fuel and production of biogas through anaerobic digestion of sugar industry wastes.

Biogas production, either from residues from the sugarcane industry, or from distillery musts, can contribute to considerable energy savings, with additional benefits related to a decrease in contamination. In Cuba and Brazil this alternative has been used on a commercial scale. The residues from the biogas production process—liquid effluents as well as solids—are of high value as fertilizers that can be returned to the soil in sugarcane areas with no ecological risk.

Bagasse as a Renewable Fiber Source

Among fibrous materials, *bagasse* has the advantage that its collection is a given of the sugarcane industry. The only problem lies in the added transportation costs from the long distances between pulp plants and sugarcane mills. Because of the present and future importance of *bagasse* as a fibrous raw material in Cuba, serious research and development efforts have been dedicated to improving these technologies. Supporting these efforts and understanding the role of *bagasse* is very important for a great number of developing countries. The United Nations Industrial Development Organization (UNIDO) has provided both cooperation and financial assistance to Cuba in this effort. With UNIDO's help, the Bagasse Cellulose Research and Production Union, also known as the Cuba-9 Project, was created. The concepts and technologies mentioned in this document are, to a great extent, the result of the work of this institution (GEPLACEA 1990).

In Cuba, harmful residues have been studied and solutions have been found. The processes of delignification and alkaline extraction with oxygen, which reduce by half the generation of chloride compounds, have been developed. The fine tuning of the production process has decreased the chemical demand for oxygen by 90 percent, and solids by 80 percent.

A high-yielding pulp production process has also been developed (Bambanaste 1988; García 1988) that dissolves only 15–20 percent of the *bagasse*, as compared with the 45–55 percent using conventional chemical technology. The levels of oxygen are half of that required in the chemical process. A high-yielding plant of 200 tons has the same ecological impact as a chemical pulp plant of 44 tons per day without recovery systems. The technical and economic feasibility for newsprint production, corrugated cardboard for boxes, and other materials have been demonstrated and internationally acknowledged (GEPLACEA 1990).

The contamination generated by residues from *bagasse* pulp and paper production can be managed well within the limits established for the wood industry, if existing technological solutions, as well as the new technologies are used (such as high-yielding pulp production). The increasing societal need for fiber by-products, together with the urgent need to preserve our forest reserves, make sugarcane *bagasse* an alternative of tremendous economic and ecological relevance.

Conclusion

Our goal has been to show that the sugarcane industry in Cuba has all the necessary ingredients of sustainable industrial development: it is economically viable, provides its own energy source, uses local raw materials, and is ecologically friendly.

"Natural" or organic sugarcane production is possible with minimal chemical inputs and little environmental pollution. Many such techniques are already used on a commercial scale in Cuba. To achieve the full potential, new scientific advances in sustainable technologies must be embraced, along with a stronger integration between field and industry.

Harvesting sugar without burning is of great importance for environmental protection. This is not only technically and economically feasible, but also introduces several additional benefits associated with the use of the no-longer-burned agricultural residues. The Cuban harvest system was designed with this concept in mind, and we have been gradually improving harvesting techniques over years of extensive operation.

Sugarcane diversification, beyond representing an alternative to the sugar market crisis, is a logical and economically advantageous development strategy. The number of highly valuable products that can be obtained from sugarcane, or from sugar industry by-products, is practically limitless.

The potential of sugarcane as a renewable energy resource is superior to that of other known crops or plant species. Sugar or alcohol production can be carried out without external energy consumption, while leaving considerable energy surpluses for other uses.

Sugarcane and the by-products of the sugar industry create feed for livestock production under tropical conditions, comparable (and in some instances superior) to the conventional pasture and grain methods imported from temperate regions.

Sugarcane *bagasse* is a good-quality fiber for the production of a great assortment of products based on pulp, paper, particle board, and chemical by-products of cellulose. In light of its renewable nature, and the available technologies, the use of *bagasse* represents an important ecological contribution as an alternative to forest exploitation.

The agricultural nature of the sugar processing industry provides an answer to what to do with its liquid residues. The cost-effective solution is recycling by returning them to the fields as fertilizer. The combination of all of these technologies in integrated systems is allowing us to work toward a truly "clean" industrial process.

References

Acosta, R. 1992. La caña de azúcar: una biomasa efectiva para disminuir las emisiones netas de CO2. IPCC/AFOS Workshop, Canberra, Australia.

Alexander, A.G. 1985. The energy cane alternative. *Sugar Series* No. 6. New York: Elsevier Science.

Alexander, A.G. 1986. Producción de caña para energía. *Memorias del Seminario Interame-ricano de la Caña de Azúcar.* Miami.

Alvarez, A. 1995. Conservación de suelos ondulados en el cultivo de la caña de azúcar. *Cañaveral,* Vol. 1, No. 4: 24.

Alvarez, F.J. 1986. Experiencias con la caña de azúcar integral en la alimentación animal en México (FAO expert consultation on sugarcane as feed). Santo Domingo, Dominican Republic.

Arrascaeta, A. 1988. Introducción a la práctica industrial del secador neumático, ICINAZ. *Revista ATAC* 45:6.

Bambanaste, R. 1988. Pulpeo quimi-mecánico de bagazo. Proyecto Cuba. IX Seminario Internacional de los Derivados de la caña de azúcar. Havana.

Bouza, H., G. Serba, R. Villegas, C. Ronzoni, S. Hernández, J. Martínez, and E. Berra. 1996. Nueva tecnología de labranza mínima en la CPA cañera "Amistad Cuba-Laos." *Cañaveral,* Vol. 2, No. 2:4.

Casanova, E. 1982. *Eficiencia agroindustrial azucarera.* Havana: Editorial Científico-Técnicas.

Cordovés, M. 1978. Nuevos azúcares competidores: Respuesta de la industria azucarera-cañera. Seminario Internacional de los Derivados de la Caña de Azúcar. Havana.

Correia, L.E. 1991. Cogeneración y producción de energía eléctrica en usinas de azúcar y destilerías en Brasil. GEPLACEA.

Crespo, R. 1996. Cultivos asociados a la caña de azúcar. *Cuba & Caña* 2.

de Armas, C. and L. Gonzáles. 1986. La caña de azúcar como fuente de energía. *La industria de los derividos de la caña de azúcar.* Havana: Editorial Científico-Técnica.

Díaz-Casas, F. 1996. Manejo integrado de malezas en caña de azúcar. *Cuba & Caña.*

Elias, A. 1986. Commercial application of molasses feeding to ruminants in Cuba (FAO expert consultation on sugarcane as feed). Santo Domingo, Dominican Republic.

Figueroa, V. and J. Ly. 1990. Alimentación porcina no convencional. Mexico City: Colección GEPLACEA.

García, O. L. 1988. Desarrollo y perspectivas de la tecnología cubana de papel periódico. Proyecto Cuba-9. Seminario Internacional de los Derivados de la Caña de Azúcar. Havana.

GEPLACEA. 1988. La industria alcoquímica en América Latina y el Caribe. Mexico City.

GEPLACEA. 1990. Proceedings del Seminario Internacional de Papel Periódico de Bagazo. Havana.

González, L. 1987. La producción de proteínas microbiales a partir de la agroindustria azucarera. Seminario Latinoamericano sobre Biotecnología. Antigua, Guatemala.

González, L. and T. Saez. 1991. La viabilidad de proyectos de desarrollo de alimento animal a partir de los subproductos de la agroindustria azucarera. Congreso ATALAC. Mexico City.

Guerra, R. 1961. *Azúcar y Población en las Antillas.* Havana: Imprenta Nacional.

ICIDCA-UNIDO-GEPLACEA. 1985. *Manual de los Derivados de la Caña de Azúcar.* Havana.

INCA. 1996. Servicio de Recomendaciones de Fertilizantes y enmiendas. Havana: Departamento de Suelos y Fertilizantes del INCA.

López, P. and C. de Armas. 1980. La potencialidad de la caña como recurso energético renovable. Seminario de Racionalización Energética de la Industria Azucarera. Havana: UNIDO-OLADE.

Ly, J. 1989. Procesos digestivos y empleo de mieles de caña para el cerdo. La Melaza como Recurso Alimenticio para Producción Animal. Mexico City: GEPLACEA.

Moreno-Fraginalis, M. 1978. *El Ingenio.* Havana: Editorial de Ciencias Sociales, pp. 91–92.

Murgueitio, E. 1990. La caña integral en la alimentación de rumiantes. *Sistemas Alternativos para Alimentación.* Mexico City: GEPLACEA, pg. 81.

Noa, H. 1982. *La diversificación de la agroindustria de la caña de azúcar.* Mexico City: GEPLACEA.

Ogden, J.M. 1990. Steam economy and cogeneration in cane sugar factories. *International Sugar* J:92:1099.

Pérez, R. 1986. The use of molasses for monogastrics (FAO expert consultation on sugarcane as feed). Santo Domingo, Dominican Republic.

Pimentel, C. 1997. Sostenibilidad de los Sistemas Pecuarios Industrializados. Taller FAO Hacia una agricultura tropic con menos uso de energía fósil. Havana.

Preston, T.R. 1977. Nutritive value of sugarcane for ruminants. *Tropical Animal Production* 2:125–142.

Preston, T.R. 1986. Sugarcane in animal feeding: An overview (FAO expert consultation on sugarcane as feed). Santo Domingo, Dominican Republic.

Preston, T.R. 1988. El ajuste de los sistemas de producción pecuaria a los recursos disponibles. Seminario producción pecuaria tropical. Cali, Colombia.

Preston, T.R. 1997. Sistemas integrados para pequeños Agricultores en el Sudeste Asiático. Taller FAO Hacia una agricultura tropical con menos uso de energía fósil. Havana.

Preston, T.R. and E. Murgueitio. 1988. La caña de azúcar como base de la producción pecuaria en el trópico. Seminario Internacional sobre derivados de la caña de azúcar. May. Havana.

Rhodes, R. and B. Rank. 1991. Algunos factores que influyen sobre la fotosíntesis de la caña de azúcar. *Ciencias Biológicas*. September.

Suárez Rivacoba, R. 1987. Experiencias y desarrollos cubanos en la producción de energía y alimento a partir de la caña de azúcar. *Uso alternativo de la caña de azúcar para energía y alimento.* Mexico City: GEPLACEA, pg. 367.

CHAPTER SIXTEEN

Case Studies: The Mixed Experiences of Two New Cooperatives

Niurka Pérez and Dayma Echevarría, Rural Studies Team, University of Havana

The recent reorganization of Cuban agriculture involved the creation of the Basic Units of Cooperative Production (UBPCs) in September 1993 (Pérez and Torres 1998). The UBPCs were created on state lands thanks to the initiative of the Political Bureau of the Communist Party of Cuba. The Ministry of Agriculture (MINAG) and the Ministry of Sugar (MINAZ) worked out the legislation needed to put this idea into practice. This reorganization of the former state farms has decisively affected how different crops are worked and processed in Cuba.

The objectives of the reorganization of land tenure and use were focused toward diversifying production technologies and actors, and a re-scaling of land use. In many cases, the search for economic sustainability and the lack of resources have led to a return to more traditional and ecological practices. Overall, the new circumstances ensured that low-input ecological agriculture be developed, and made all too clear the need to find new stimuli to stabilize the labor force in the rural sector (Figueroa 1998).

In this chapter we summarize the experiences, compiled from extensive interviews, of two typical UBPCs, one in tobacco and one in sugarcane.

Creation of the UBPCs

The principal forerunners of the UBPCs—those who laid the groundwork for cooperative and collective production in Cuba—were the Sugarcane Cooperative Units created in the 1960s, and the Agricultural Societies and Agricultural Production Cooperatives (CPAs) set up in 1975 (Echevarría 1996).

When the UBPCs were set up, Law 142 was passed, stating that: "... other incentives should be found that motivate producers to deliver their productive assets in order to achieve the greatest production at the lowest possible resource cost. Therefore, it is necessary to carry out important innovations in state agriculture, such as the creation of Basic Units of Cooperative Production as a new form of agricultural-livestock production organization" (MINAG 1993).

The new enterprises would be governed by four main principles set forth in the legislation: 1) foment a new relationship between the producer and the land in a way that would stimulate the farmer's commitment, and also his/her individual and collective responsibility; 2) on-farm self-provisioning of basic foodstuffs for the workers and their families, who work cooperatively toward the gradual improvement of their living conditions; 3) worker's profits to be based on production; 4) development of management autonomy and the self-administration of resources, so as to become self-sufficient over time.

The rhythm of transformation from state farms to UBPCs has varied by sector. Where sugarcane was the main crop, there was rapid progress. In non-cane sectors it has been more gradual.[1] In general, the design of the UBPC as an economic unit did not take into consideration the diversity among crops, regions, microeconomic relationships, or the characteristics of the individual units (Pérez et al. 1996).

A UBPC is defined, in the general regulations, as "an economic and social organization composed of self-sufficient workers in the productive field receiving land in usufruct, for an unlimited period of time, and having their own legal status" (MINAG 1993). Other characteristics of this new form of property include the fact that the members participate voluntarily, and they collectively own the production that is sold—to a state enterprise, or on the open market, in the case of surpluses—which determines the form in which it should be sold.[2] The members must pay for the inputs they use, and the basic equipment they received with the land (in the form of a loan paid over time).[3] They elect their administrative board, which must periodically account to the members of the General Assembly, and ultimately must approve fundamental decisions.

Characteristics of the UBPCs

After more than five years of membership in UBPCs, the workers have faced important changes in their lives. This is a product of all four governing principles of the UBPCs, but especially of the development of self-management, and the new relationship of the producer to the land.[4] It is one thing to be a salaried farm worker, whose salary is fixed regardless of production, and quite another to be a farmer-member of a self-managed cooperative, whose profits depend entirely on production.

At a macro level, the most significant advances of the UBPCs to date, in comparison to the old state farms, are:

- greater interest by workers in production costs, levels of production, and profits of the enterprise, and their participation in the administration boards
- reduction of losses due to widespread cost-cutting
- new opportunities for participation in the production process and in marketing, particularly via sales at farmers' markets
- tendency to adopt agroecological practices to compensate for the lack of inputs resulting from the collapse of trading relations with the former Socialist Bloc countries.
- tendency for UBPCs to converge on optimal sizes and management practices according to the type of crop and land being farmed (Pérez et al. 1998)

This can be seen as a transition, forced by circumstances, from conventional agriculture based on the use of chemical fertilizers, pesticides, mechanization, irrigation, and petroleum, to an alternative model that involves organic or alternative farming practices. The new model promotes organic fertilizers, biological pest control, animal traction and other alternative energy sources, crop and pasture rotations, intercropping, and rediscovering knowledge of traditional farming practices among the rural population (Zaragoza and Gloria n.d.).

The cases presented here demonstrate the variability in the degree to which this transition has happened in the young UBPCs, depending in part on the type of crop being produced. An important caveat is that the process is much more advanced among small farmers and members of the older forms of cooperatives, than on the UBPCs.

Experiences of Tobacco and Sugarcane UBPCs

Tobacco has always been important as an export crop in Cuba. It is mainly grown in the central and eastern part of the country, though it is grown for local consumption in other areas. Tobacco production had been dropping for years due to out-migration of farmers and the conversion of land to other crops. But since 1993 it has been in a process of rapid recovery due to a shift in policy. Tobacco UBPCs were set up, and land was granted to individual producers who agreed to grow tobacco,[5] creating a dynamic sector where private producers and cooperatives have been leading the way. In 1998, of a total of 2,701 UBPCs, 53 engage in tobacco production (MINAG 1998).

Now a dynamic crop, tobacco receives foreign investment; it receives more productive resources and contributes the most foreign exchange to MINAG of any crop. Tobacco profits are then used to reactivate other crops of national importance. In the same farm where tobacco is grown, other crops are produced for family consumption and to sell at the markets.

Sugarcane production is a national agroindustrial activity that can be found in all of Cuba's provinces. Historically it has been the main national crop, and is the number one employer in Cuba, with both agricultural and industrial workers. Cuba has 154 Agro-Industrial Complexes (CAI) for sugarcane; as of September 1997, there were 1,063 sugarcane UBPCs with 141,785 members (MINAZ 1997).

La Jocuma UBPC

The La Jocuma UBPC is located in the Consolación del Sur municipality (in Pinar del Río province). The members have a strong tradition of growing tobacco. The UBPC was created in September 1993 on the Juan Casanueva state farm, belonging to the Consolación del Sur State Tobacco Enterprise.[6]

The members have partially returned to the practices of the *vega campesina* (traditional small tobacco farm). These practices include intensive cultivation of food crops on all soils not considered ideal for tobacco, and intercropping the tobacco with secondary crops that do not affect its yield or quality. On the UBPC, cassava, sweet potato, taro, and to a lesser extent, cabbage, lettuce, radish, and tomatoes are grown. Overall, the lack of inputs has fostered the return to practices and traditional technologies seldom seen at state farms.

The tobacco parcels are rotated with maize. They are not rotated with sweet potato, because of pest problems and because it takes up the same nutrients as tobacco. The parcels cannot be rotated with beans either, since they are frequently attacked by blue mold.[7] Corn, however, is frequently intercropped with the beans.

Tobacco seed is obtained through the Empresa Tabacalera, which in turn purchases it from the Ministry of Agriculture seed enterprise (Empresa de Semillas Varias del MINAG). In 1994 and 1995 the tobacco enterprise produced its own seed, but production was discontinued due to high costs. The tobacco variety to be sown is determined by the UBPC enterprise. When the UBPCs were created they planted the Burley variety, typified by its high yield and resistance to blue mold, but it was frequently attacked by other fungi (Espino et al. 1998). Later they planted the Habana 92 variety, which has similar characteristics but lower yield, needs more care, and is more resistant. During the 1997–1998 season the Burley variety was sown again but was affected by the highly damaging *Fusarium* pathogen. In 1998–1999 Habana 92 was planted.[8] The UBPC administrative board is concerned with the constant change of varieties.

Seedling production has been done on-farm since the 1996–1997 season. These seedlings have a higher survival rate than those purchased—because they are much fresher when they get to the sowing areas—although they need adequate care and inputs. In an effort to diversify their sources of income, the UBPC now sells seedlings to individual farmers in the surrounding areas.[9]

The seed for other crops is also produced at the UBPC. Different methods are used to preserve the seeds according to their type. For example, bean seeds for sowing are preselected, protected from light and air, and stored in a wax-sealed plastic tank. After selection, corn seeds are stored in grain form or on the ear; later they are selected for quality, and then treated with a powder to deter pests.

Traditional methods of improving eroded soil are being used again before soil preparation; the areas are rotated every year. In improving the soil, farm workers spread out dry tobacco stalks that are left over after leaf harvest; they also gather fertile soil brought back from silted rivers or stream beds; and use earthworm humus, chopped corn stalks, etc. This is done one or two months prior to soil preparation. Very occasionally they also use granulated chemical fertilizers applied by hand. Herbicides are only used in the bed were seedlings are produced for transplant, not in the open field.

Weeding is generally done by hand with a hoe; sometimes a weed chopping machine is used. In 1992 and 1993 all the areas to be sown were weeded by a special weeding brigade. The timber found on new areas cleared for cultivation was made into charcoal.[10]

Animal traction is generally used for soil preparation. Farmers break the soil with a furrow plow, then depending on fuel availability, use oxen or a tractor with a chopping machine once or twice, and then the soil is disc harrowed (the implement has two slanted blades, different from the one-bladed American plow).

The tobacco seedlings are then hand planted and irrigated. After being transplanted, the plants are irrigated at seven days, and resown where necessary. Fertilizer is then applied and the plant is hilled over.[11] Two passes are made with a tiller at 18–20 days, covering the base of each plant with earth, and it is irrigated again. The rest of the labor is done by hand. This includes hoeing, spraying (every 7, 14, and 21 days), a once-over inspection (the buds in the intersection of leaf and stem are eliminated), harvest of larger leaves 45 days after sowing, and threading, a task usually carried about by women that consists of sewing tobacco leaves in pairs and placing them on horizontal supports for drying. Sprouts are again removed from the intersection of leaf and stem of the plant. Reharvesting takes place seven days after the first harvest, as well as harvesting of the crown, threading, and cutting the lower-quality tobacco leaves.

The leaves can be collected alone or by cutting sections of the plant; this should be done 65 to 70 days after sowing. Leaves are placed in tobacco houses and dried for one to one and a half months. The threads are later removed and the leaves placed in heaps for 20 to 25 days (although Burley tobacco does not use this process) (Espino et al. 1998). Later the tobacco is baled and selected for sale.

Traditional methods are generally used for other crops. Animal traction is used for soil preparation and labor is done by hand or with oxen. No chemical fertilizers are applied, while organic fertilizers are of great help.

Pests are controlled by chemical and biological methods in tobacco. Pesticides are not applied to the rest of the crops. One pesticide used is a substance called *tabaquina* which the UBPC produces. To avoid the attack of *Thrips palmi*[12] and other common tobacco pests, a trap is made of bamboo poles and white plastic bags coated with molasses residues, placed a half meter from the row. The insects are trapped in the bags. Corn is also planted around the area as a barrier for pests.[13]

Watering is done by a gravity irrigation system using motors that pump water from the surrounding rivers and streams. Seedlings require a lot of moisture, hence irrigation is very important. Depending on the degree of soil moisture, tobacco is irrigated weekly up to the first harvest. When irrigating in the furrows, the flow of water is directed to the driest area at regular intervals, with the assistance of hoes. Other crops use the same system depending on fuel availability for the irrigation equipment. One of the persons interviewed said, "no water is stored when it rains, this is only done by the peasants, but not here."[14] Harvest residues are used, however, to feed pigs and poultry. Beef, sheep, and goats graze on natural pastures.

Carlos de la Rosa UBPC[15]

The Carlos de la Rosa sugarcane UBPC of Havana province belongs to the Osvaldo Sánchez Agro-Industrial Complex, located in the southeastern part of Guines municipality. It has a total area of 1,472 hectares of which 1,244 are dedicated to sugarcane cultivation. In the 1998 harvest, 1,119 hectares were in sugarcane, and 920 hectares were cut. The self-provisioning area remained constant at 38 hectares.

The Jaronú 60-5 variety, characterized by its high sugar content but poor resistance to smut disease, was historically sown in this zone. This variety was later replaced by Barbados 43-62 and Jaronu 43-72, both with good yields, but also sensitive to the smut fungus. There are more than 13 varieties in Cuba affected by this disease.

In the UBPC the best sugarcane stands started to deteriorate. This process began in 1990 and became more acute from 1993 to 1995, as no fertilizer was applied due to the economic crisis. In more recent sugarcane harvests, fertilization was again done in the sugarcane areas, together with weeding and earthing over, to attempt to boost yields.

In 1995, strains were reintroduced that allow for a more mature sugarcane, which is essential for higher yields per area. Intercropping in sugarcane was done only once with beans. Crop association is not done in non-sugarcane commercial crops on this UBPC, only crop rotation with rice being planted where the sugarcane once grew. Other crop rotation and cattle raising are not practiced on this cooperative. In food production areas, however, intercropping is done with squash/corn, beans/corn, and tomato/papaya.

During the first years of the Special Period, sugarcane production and especially seed banks were affected. Seed bank activity was limited at the UBPC, but not abandoned. The UBPC has a good certified seed bank,[16] which comes from the CAI's registered seed bank, which in turn obtains supplies from the Experiment Station where they undergo a careful selection process.

Weekly phytosanitary inspections are carried out in the UBPC seed bank, by technicians from the Registered Seed Center. Formal inspections by the certifier occur every three to six months.

The Jaronú 60-5 variety showed the highest yield in the 1997–1998 harvests, but with a high incidence of smut fungus. Sanitary measures to combat this fungus are carried out manually, and now less frequently, as planting is increasing, an expanded labor force is needed and only a few brigades are available for this activity. All the varieties now planted are high yielding, especially Tayabito.[17]

Mechanization is used for soil preparation in all of this UBPC (oxen are rarely used). The reasons for this, according to the people interviewed, are that it is hard to get the oxen to the distant places (the area of the cooperative is approximately 1,248 hectares), and there is no place in such areas where the animals can be kept once the work is completed.

To prepare soils an interval of 10 to 15 days is needed between the first and second pass with machinery, so the weeds can dry and their seeds do not germinate (the seeds are exposed to the sun in order to dry out). This is an efficient practice for weed control and leaves the soil less compacted when plowing.

The soil preparation steps are as follows: a second harrowing, heavy and light harrowing, going over the fields with the leveller, deep plowing, and preparation and planting (Álvarez 1995). The program and sequence of these operations are similar in winter and spring plantings and are only modified when planting in colder weather or in the springtime, when rain is a factor. Camilo Ramos Ramos, an experienced worker of the Carlos de la Rosa UBPC interviewed in 1998 by Miriam García, reports that sowing is very demanding at the UBPC: though the method now employed is very effective and almost 100 percent of the sugarcane germinates, it is very expensive and all the activities must be done manually.

Sugarcane cutting for propagation is one of the first steps in planting; it is done in May and can last 45 days, depending on the extent of the land programmed for sowing and the work force available.[18] Sugarcane is manually winnowed so that the stalk buds may not be damaged. After this, the stalk is chopped perpendicularly in pieces containing three to four buds, with one strike of the machete on a rubber or wooden surface. After this, the plant is placed in a cart and taken to the field to be planted.

The head of the UBPC brigade, who has ample experience in agriculture, was interviewed and explained that weeding should not be done unless it is clear that the weeds present are damaging to the crop (Rodríguez and Espino 1996), and depending on the type of soil, moisture content and environmental conditions.[19] He also pointed out that weeds should not just be controlled, but rather should be well managed, based on the floral composition and periods when the weed community really interferes with the crop. Also, a combination of diverse operations, measures, and conservation methods should be adopted to control weeds such as cultivating between rows, weeding, and limited herbicide use in order to achieve more cost-effective management.

The weeding plan depends on weed density. The persons interviewed confirmed that the more tillage operations, the more effective the control of weeds. Other preventive measures can include weeding of the margins of the fields, fences, and roads.

The administrator of the UBPC, José Luis Hernández Riol, stated that fertilization must not be done unless there is a high probability that it will really increase yields and be profitable.[20]

Mercedes Oliva, an agronomist working at the UBPC, explained the different steps to establish fertilizer indices in each sugarcane area.[21] After the second plowing and weeding, ammonium and urea are applied, which provokes a high productive response. In young sugarcane plantings, phosphorous and potassium should be used; to do this the soil is plowed and fertilized as near as possible to the plant so that the fertilizer gets into the earth and then is covered with soil. After the harvest, the sprouts are cultivated and ammonium and urea fertilization are applied. Twenty or thirty days later, the plants are watered; and if this is not done, the fertilizer will be lost.

Chemical pest control is not used. The biological control program involves use of the parasitic fly *Lixophaga diatraea*[22] against the sugarcane borer, *Diatraea saccharalis* (Guzmán and Pablos 1996). Generally, it has been released before the sugarcane harvest or after, if it is available at that moment in the CAI. The standard that the UBPC uses is 1000 flies over 13 hectares in fields for sugarcane seed production, and 500 flies for general sugarcane production, and this has been very effective.

The material harvested at the UBPC is sampled to determine the damage caused by smut (*Ustilago scitaminea*) and rust (*Puccinia melanocephala*).

Rodents also affect sugarcane, although less than the sugarcane borer (Padrón Mestre 1996). They can affect the sugar content and weight due to the fermentation of sugar in the damaged stalks and increased susceptibility to diseases, mainly red rust. Among the measures to eliminate rodents are weeding, elimination of wastes, and the use of traps. Recently, the Cuban biopesticide *Biorat* has been used as a biological control agent.

Sugarcane harvests are programmed after the sugarcane areas are sampled and then tested at the CAI laboratories to determine yield, maturity, and purity. Depending on the results, the variety and age of the plants will determine when the harvest will be done. The plants selected should not be less than 18 months old, having reached an 85 percent maturity level.

Cuban KPT2M and KPT machinery is used in the harvesting process. Irrigation depends on rainfall and the vegetative development of the variety.

The people interviewed deemed that irrigation is limited by fuel availability, thereby affecting sugarcane yields (Ramírez Miguez 1995). At this UBPC, irrigation is hardly used at all. When done at all, gravity channels are used with a system of wells and motors or turbines. These wells are distributed in the fields.[23] When the CAIs are processing the sugarcane, the residual waters irrigate 80 hectares of the cane cooperative. Many years ago, the cane was irrigated with waste water from the Quivicán municipality, but studies demonstrated that the aquifer was contaminated and its use was prohibited.

Conclusion

It is clear from these two contrasting cases that the historical characteristics of each crop produce considerable differences in ecological practices in the young UBPCs. Sugarcane is a highly industrialized crop, produced on large areas where machinery is used to make the work more humane (in spite of the fact that it may affect the soil). Tobacco is an intensive, hand-labor, family crop, which is produced in small areas and has a semi-artisanal production cycle with a minimum utilization of machinery.

During the 1970s through links with the Socialist Bloc in Europe, a model emphasizing the application of chemical products and heavy machinery was implemented. Sugarcane production was the first priority because of its economic importance for Cuba. When the Socialist Bloc disappeared, sugarcane production was severely affected, and has yet to recover.

Because tobacco producers always used some traditional ecological methods, the lack of inputs caused less dramatic problems. The search for ecological or non-ecological alternatives continues in the sugarcane UBPC, because of their dependence on industrial inputs and machinery.

The truth is that there is still a long way to go for the development of organic production at the large farm level, in both sugarcane and tobacco crops, although at the small-scale individual and cooperative level many farmers are very far along the road to achieving this goal. It should be kept in mind that the UBPC members have only recently become farmers, while individual *campesinos* and members of longer-standing types of cooperatives have years—and indeed generations—of experience with farming practices.

Notes

1. About 87.3 percent of the land in the sugarcane Agro-Industrial Complexes (CAIs) was turned into 1,576 UBPCs (MINAG 1993, pg. 25).

2. This enterprise is the regulatory and commercial entity of the UBPCs.

3. At the beginning the means of production of the UBPCs were state property and belonged to the state farms from which the UBPCs were created. Later the equipment was sold to the UBPCs on credit (five years) at low-interest rates.

4. The link between the farmer and the land mainly consists of how work is organized, and how profits are now linked with the final production results.

5. More than 13,000 individual peasants have received 27,000 hectares of land for tobacco cultivation.

6. The UBPC lands are considered the best tobacco plantations.

7. An explanation of blue mold and its consequences for tobacco crops can be found in Espino et al., pp. 92–93.

8. Interviews with Arnaldo Vigoa, administrator; Reynaldo Fernández Paulino, chief of production; and Francisco Hernández, member of the UBPC La Jocuma, March 1998, by Ernel González Mastrapa, Niurka Pérez Rojas, and Miriam García Aguiar.

9. The cost of 1,000 seedlings ranges from four to six Cuban pesos and they are sold at seven to ten pesos. Interview with Arnaldo Vigoa, administrator, March 11, 1998, by Ernel González Mastrapa.

10. Interview with Reynaldo Fernández Paulino, chief of production, March 11, 1993 by Niurka Pérez Rojas.

11. This task is carried out with the feet, hilling up earth, covering the base of the plant.

12. Espino et al., op. cit. pg. 92. It is thought that this insect was introduced by the US to sabotage certain crops.

13. Interviews with Luis González León, Pedro León Rodríguez, and Reynaldo Fernández, March 11 and 12, 1998, by Miriam García, Niurka Pérez, and Ernel González.

14. Interview with Luis González León on March 12, 1998 by Miriam García.

15. Interviews with Mercedes Oliva, agronomist; José Luis Hernández Riol, administrator of the UBPC; Jesús Martín, chief of production; Ricardo Máximo Torres, chief of transportation; Ismael Lorenzo Rosa Pérez, chief of the self-provisioning brigade; Camilo Ramos Ramos, chief of the sowing brigade; and José Ignacio Gómez Pérez, member of the cooperative. Interviews were conducted by Miriam García, Niurka Pérez, and Ernel González from March to November 1998. These cases were reported in García and Morell 1998.

16. The search for new varieties is being conducted in the UBPC. Among the main varieties are: Tayabito (Ty 86-28 and 7017) and Mayarí (My 5514). Although adapted well to these soils they have certain inconveniences, since they are long-cycle varieties that need abundant irrigation. The lack of fuel is one of the main difficulties for irrigation.

17. Estimated by the Administrative Board of the UBPC. At the end of 1999, 94 percent of these seeds were certified.

18. Three brigades are usually employed at the UBPC. Two of these brigades use from 20 to 30 men and the third from 8 to 10.

19. Interview with Ismael Lorenzo Rosa Pérez by Miriam García. August–November 1998.

20. To produce one ton of stalks, sugarcane extracts 1.5 kg N, 0.5 kg P and 1.8 kg K_2O from the soil. Federico Sulroca Domínguez. MINAZ. El ABC de los fertilizantes y su amenejo en la caña de azúcar. *Canaveral* 1:2 April–June 1995.

21. The following steps are taken for sugarcane fertilization: a bimonthly chemical sampling of the soils, with analysis of the samples in specialized laboratories such as Fertilizers and Varieties Service (SERVAR). Calculation of the fertilizer deficit and dosages (laboratory results). Data and field maps to establish cartograms with the fertilization indices of each sugarcane field. The UBPC uses two ammonium fertilizer machines, four F350 fertilizer spreading machines, and four MTZ-80 tractors. Fertilization by plane is not usually used.

22. Before the 1980s, the Ministry of Sugar established artisanal rustic biocontrol laboratories to rear this parasitic fly.

23. The wells are located in the following areas: Los Mangos, Guásimas, Prado, Ojo de Agua, Los Domínguez, and Cayo Alto. Those of the Rice Plan now belong to the Sugarcane Plan. There are 10 wells with their respective motors and turbines. Although the Mayabeque river surrounds the UBPC, the waters are diverted through a channel and cannot be used for irrigation.

References

Álvarez, A. 1995. Por qué es importante una correcta preparación de suelos en la caña de azúcar. *Cañaveral*, Vol. 1, No. 2, pp. 23–25.

Echevarría, D. 1996. Relaciones de las UBPC tabacaleras con sus miembros y con la empresa estatal. Estudios de caso en Consolación del Sur (thesis). Havana: University of Havana.

Espino, E., V. Andino, and G. Quintana. 1998. *Instructivo técnico para el cultivo del tabaco.* Havana: SEDAGRI-AGRINFOR, MINAG.

Figueroa, V. 1998. El nuevo modelo agrario bajo los marcos de la Reforma Económica. *Desarrollo rural y participacíon* 1–45.

García, M. and R. Morell. 1998. Informe preliminar del Proyecto de Producción de Azúcar Orgánica. Havana: University of Havana.

Guzmán, A. and Pablos, P. 1996. Bórer, raquitismo de los retoños y nutrientes en los suelos. *Cañaveral* 3:4 October–December, pp. 26–27.

MINAG. 1993. *Legislación sobre las UBPCs.* Havana: MINAG, pp. 1, 3–4.

MINAG. 1998. Economic results of UBPCs and CPAs by territories, branches, and at the national level. Havana: MINAG.

MINAZ. 1997. UBPCs by province, from September 1997. Havana: MINAZ.

Padrón Mestre, V. 1996. Los roedores dañinos en la caña de azúcar. *Cañaveral* 2:2. April–June, pg. 42.

Pérez, N., E. González, C. Torres, and M. García. 1996. Cambio en el agro urbano: la experiencia de las UBPC cañeras en la agricultura cubana a través de estudios de caso. Equipo de Estudios Rurales. Havana: University of Havana.

Pérez, N. and D. Echevarría. 1997. Participación y producción agraria en Cuba: las UBPC. *TEMAS: Cultura, Ideología, Sociedad* No. 11, pp. 69–75.

Pérez, N., G. Mastrapa, E., and M. García. 1998. La transformación de la agricultura cubana a partir de 1993. Havana: University of Havana.

Pérez, N. and C. Torres. 1998. Las UBPC hacia un nuevo proyecto de participación. Desarrollo rural y participación. Equipo de Estudios Rurales. Havana: University of Havana, pp. 46–67.

Ramírez Miguez, J. 1995. ¿Porqúe un riego eficiente? *Cañaveral* 1:2 April–June: 14.

Rodríguez, C.N. and A. Espino. 1996. Guerra inteligente a las malas hierbas. *Cañaveral* April–June: 55–57

Zaragoza, Z. and A. Gloria. n. d. Uso de la tierra y diferentes formas de organización de la producción cañera en el municipio Guines (thesis). Havana: University of Havana, pg. 73.

EPILOGUE

The Unique Pathway of Cuban Development

Richard Levins, Harvard University School of Public Health

The world is grappling with a terrible dilemma: on the one hand the world's peoples and most governments aspire to a rising standard of living. On the other hand, if that rising standard of living is taken to mean following the pathway of development and attaining the life style of Euro-North America, it would destroy the life support system of all of us. The vast majority of the world's population can neither be kept in poverty nor be drawn into the consumerist way of life. The goals of justice and sustainability seem to be mutually exclusive. But when two equally urgent goals are incompatible, it is often the case that we have posed the problem too small. As long as development and progress are seen as taking place along a single pathway from less developed to more developed, there is no solution. But what if development is seen as a branching pathway, with choices all along the way? What if a rising standard of living is seen mostly as a rising quality of life within the constraints of sustainability? What if economic development is not a goal in itself but a means to enriching life and preserving nature, with emphasis on equity, health, education, culture, recreation, and mutual caring within an environment which is sustainable, diverse, and people-friendly? That is the unique pathway that Cuba has embarked on.

Every kind of society has its own pattern of relations of people with each other, with the production of life's necessities and with the rest of nature. Every major change in social form has carried with it changed relations with animate and inanimate nature—with plants and animals, with pathogens, with soil, water, and air. What we see in Cuba is the emergence of a socialist society, with socialist patterns of production, demography, settlement, human relations, and relations with our habitat. Agriculture is one aspect of this process.

These changed relations develop unevenly in the course of getting by, solving problems and thinking about problems on the run. Nobody could really sit back and ask, what should we do to have a socialist landscape, or forest or crops or technology. And yet, that is what is happening. Cuba is creating a socialist ecology, a socialist pattern of relations between humans and the land, the insects, the fungi, and tomatoes. Sometimes it seems as if the changes are random as ideas are tried, abandoned, tried again in new ways. Sometimes it seems as if innovations are merely emergency responses to the embargo and the collapse of the Soviet Bloc. And the people most immersed in the process have had little time to stand back from their works and say, "look, everybody, we're creating a socialist eco-social system!"

But that is what is happening. And that is what is missed when visitors admire particular innovations and hope to adopt some of them in their own countries: Cuban agriculture is a socialist agriculture, created under difficult conditions to meet socialist goals.

What, then is "socialist" about Cuban agriculture?

Under socialism, ownership is in the hands of the "associated producers"—either directly in cooperatives or indirectly through the state. Where management by the state proved too indirect, inflexible, and not sufficiently democratic, new forms of ownership were adopted. Farmers' organizations and local government often lead particular production programs. The various forms of rural democracy are needed not only for making the general policy decisions of the enterprise, but also for participating in nation-wide decisions and mobilizing the collective intelligence of farm workers in solving problems jointly with scientists.

Under capitalism we have the maximum plunder of the countryside by the rulers of the cities. This is redressed under Cuban socialism in several ways:

- Priority investment in the countryside and small towns, with guaranteed equal educational and health service opportunities. Often the most advanced technologies go first to the isolated areas in the mountains, such as photovoltaic solar energy collectors;
- Recycling of farm waste within the farm and the return of urban waste to the farms, which requires that these be non-toxic;
- Recruitment of city people to move to the countryside or to work there periodically;
- Development of urban and suburban agriculture, and of rural industry;
- Research aimed at meeting socially determined needs.

First, decision making power and ownership are in the hands of the "associated producers,"—either directly in the various forms of cooperatives or indirectly through the socialist state. There is no single form of land tenure that is required by socialist theory. The transfer of land management from state farms to cooperatives was no abandonment of socialism but a reorganization within socialism to meet socialist goals better. What is precluded is privately owned agribusiness, hiring wage labor to extract maximum profit, selection of crops for their market value only, and of technologies that undermine the health of farm workers and the future capacity to produce. It also precludes a profit-driven agricultural inputs industry. This changes the priorities for research, the criteria for investment and the decisions about what and how to plant.

The top priority is feeding the whole population, and directing the products of agriculture where they are most needed. The principle that everybody eats meant that as long as milk was in short supply it went to children and was not turned into high-value derivatives such as butter and cheese. (I recall the pitifully thin pizzas of 1968 when drought reduced milk production). It meant that research was directed toward the crops that people consumed, and which were often sold to them below the costs of production. The questions of land use, of the division of labor among rural, suburban, and urban agriculture, are solved by considerations of social need, including production and employment.

A second priority is economic solvency. But this is not the same as profit maximization. Land is devoted to export crops to the extent that foreign exchange is needed to support development and production, and to supplement what is needed by the population. Nor can economic solvency be achieved at the expense of the deterioration of the soil, loss of biodiversity, or contributing to global warming. One way of increasing the economic viability of agriculture is reducing the use of off-farm inputs. Composting, recycling within the farm, biological pest management, and use of biofertilizers fit into this priority. An option that is not available is downsizing, creating unemployment in order to improve the bottom line. On the contrary, urban agriculture is encouraged—in part because it creates employment at the rate of about 20 jobs per hectare. Not all farms are solvent although they are moving in that direction. While the older National Association of Small Farmers (ANAP) cooperatives are stable economically, only about half the newer Basic Units of Cooperative Production (UBPCs) are already profitable. But given that the new cooperatives are a developing in a socialist context, solvency can be expected as the members learn to manage.

A third concern is protection of people's health. This is not an "externality" the way it is calculated under capitalism. Anything that affects the life of the population is "internal" to the society and therefore a concern. This is one of the reasons for adopting ecological methods of pest management.

The way research is done is also determined by socialist goals and socialist means. Science is not a commodity, directed toward the production of the most profitable commodities in a race for patents. Whereas under capitalism a company directs much research effort to denying that their product does any harm, Cuban research seeks out the consequences of a technology as it percolates through the environment: every consequence is important and there are no externalities. Institutions are able to work together to meet long term or urgent goals without hiding their findings as proprietary secrets. Therefore the intellectual resources of the country can be mobilized toward shared social goals. And the full access of women to positions of leadership in science doubles the brain pool available for research.

Of course this process is not easy. The prestige of Euro-North America weighs heavily on the scientists of a developing country and acts against independent innovation.

Decisions about pesticides, about specialization, about livestock technology and mechanization followed prolonged debate which is still taking place. In Cuba as in the United Sates there were debates about pesticides. But a debate about pesticides in Cuba is different from a debate about pesticides in the United States. In Cuba, as in the United States, the debate can be frustrating. We can find stubbornness, conservatism, ignorance, and even stupidity in any country. But what is different is that in Cuba the debates were expressions only of differences of opinion and therefore reason can eventually prevail. In capitalist countries, debates about technology are often weapons in the conflict of interests. The makers of pesticides never ask what might be the best way of reducing pest damage while protecting soil and people, but rather, what is the best way to turn oil into marketable commodities to sell to farmers, and they defend their products with a ferocity driven by the bottom line.

The ecological transformation of Cuban agriculture since the early 1990s is overwhelmingly complex, including changes in agrotechnology, land tenure and use, social organization of production and research, educational programs, and financial structures. From the point of view of the participants it comes about through solving problems within the context of socialist goals, goals such as stable food production in the face of natural and economic uncertainty, input reduction, preservation of productive capacity, conservation of biodiversity, protection of people, stable and healthful food supplies, recycling of residues between town and country and ruralization of cities, control by the associated producers without specifying the form, and research for social ends.

The ecological transformation of Cuban agriculture takes place in the context of ecological development as a whole—Rio, Kyoto, the Montreal protocol on ozone; programs against desertification; reforestation, protection of shorelines, integration of education, research, economics, and ecology.

We can expect the future of Cuban agriculture to be an ecological agriculture. A mosaic pattern of land use is developing in which each plot of land contributes direct products, but also contributes to the other plots: forests give lumber, charcoal, honey, and nuts, but also shelter many species of wildlife, provide refuge for the birds and bats that consume pests, regulate the flow of water (thus reducing the dependence on fossil fuels to pump irrigation water), serve as wind breaks and barriers to the spread of pests, and create special microclimates at their edges to a distance of about ten times the height of the trees. Pastures produce the usual livestock products but also retard erosion as compared to row crops, collect manure, offer a diversity of nectar sources for bees and for parasitic wasps that infect pests. Livestock graze on the weeds of fruit orchards, and so on. Farming takes place in the countryside but also in suburban and urban areas where it ameliorates the harshness of the city landscape, brings perishable foods closer to the consumers, provides jobs, and strengthens the neighborhood-sense of community.

All of this would mean a system of great complexity adapted separately to each locality by those who work there. Ecological agriculture demands the combination of physical with intellectual labor by farmers—both a general long-term socialist goal, and a way of creating jobs that could keep highly educated Cubans in the countryside.

To understand Cuban agricultural development it is first necessary to look at it closely in the richness of detail described in this volume. Then we have to step back and squint to capture the truly novel pathway of development as a whole that Cuba is pioneering. And then once again we have to focus in on the details, and glimpse the processes through which, in the course of solving particular problems, Cuba is creating something truly new and hopeful for all of humanity.

List of Acronyms

AAC Cuban Sugar Annual Report
ACAO Cuban Organic Farming Association
ACC Cuban Academy of Sciences
ACPA Cuban Animal Production Association
ACSUR Association for Cooperation with the South (Spain)
ACTAF Cuban Association of Agricultural and Forestry Technicians
AGRINFOR Agricultural Information Agency (Ministry of Agriculture)
AI artificial insemination
AIAB Italian Association for Organic Agriculture
ANAP National Association of Small Farmers
ATAC Cuban Association of Sugarcane Technicians
ATNESA Animal Traction Network for Eastern and Southern Africa
BFW Bread for the World (Germany)
CAI Agro-Industrial Complex
CAN National Poultry Complex
CCS Credit and Service Cooperative
CEAS Center for the Study of Sustainable Agriculture (Agrarian University of Havana)
CEC cation exchange coefficient
CEDECO Organization for Costa Rican Development
CEE State Committee of Statistics
CEMSA Seed Research Center
CERAI International Center for Rural and Agricultural Studies (Spain)
CIARA Foundation for Training and Innovation in Rural Development (Venezuela)
CIAT International Center for Tropical Agriculture (Colombia)
CIC Council of Churches of Cuba
CIDEA Environmental Education and Information Center
CIERI Center for the Study of Inter-American Relations
CIMAR Research Center for Oceanography and Limnology
CIMMYT International Maize and Wheat Improvement Center (Mexico)
CIPS Center for Psychological and Sociological Research
CITMA Ministry of Science, Technology, and Environment
CLADES Latin American Consortium for Agroecology and Development (Chile)
CNSV National Center for Plant Protection
COMECON Council for Mutual Economic Assistance (economic union of the former Socialist Bloc)
CPA Agricultural Production Cooperative
CREE Center for Production of Entomophages and Entomopathogens

DECAP Department of Project Coordination and Assistance (Council of Churches of Cuba)
DGSF General Directorate of Soils and Fertilizers
EDICA Animal Science Institute, Publications Department
EEPF "Indio Hatuey" Pasture and Forage Experiment Station
EER Rural Studies Team (University of Havana)
EJT Young Workers' Army
ENCC National Coffee and Cocao Experiment Station
ETPP Regional Plant Protection Station
FAO Food and Agriculture Organization (United Nations)
FAO-JAD Food and Agriculture Organization (United Nations)–Agricultural Development Council
FAR Revolutionary Armed Forces
FIDA United Nations International Fund for Agricultural Development
FLACSO-UH Latin American and Caribbean School of Social Sciences, University of Havana Campus
FMC Federation of Cuban Women
GAO Organic Farming Group of the Cuban Association of Agricultural and Forestry Technicians (ACTAF)
GENT New Type State Farm
GEPLACEA Group of Sugar Exporting Countries of Latin America and the Caribbean
GNAU National Urban Agriculture Group
GNP gross national product
GTA Regional Food Board
GVC Civilian Volunteer Group (Italy)
HIVOS Humanist Institute for Development Cooperation (The Netherlands)
IACC Civilian Aeronautics Institute of Cuba
ICA Animal Science Institute
ICRT Cuban Institute of Radio and Television
IES Institute of Ecology and Systematics
IFOAM International Federation of Organic Agriculture Movements (Germany)
IIA Rice Research Institute
IICF Citrus and Fruit Research Institute
IIHLD Liliana Dimitrova Horticultural Research Institute
IIMA Agricultural Mechanization Research Institute
IIPF Pasture and Forage Research Institute
IIS Soil Research Institute
IISA Soil and Agricultural Chemistry Research Institute
IM Institute of Meteorology

IMV Institute of Veterinary Medicine
INCA National Institute of Agricultural Sciences
INFOMUSA International Magazine on Banana and Plantain
INHA Institute for Food, Nutrition, and Hygiene
INIE Institute for Economic Research
INIFAT National Institute for Fundamental Research on Tropical Agriculture
INISAV National Plant Protection Institute
INIVIT National Research Institute for Tropical Roots and Tubers
INRA National Institute for Agrarian Reform
INRH National Institute of Hydraulic Resources
IPA Agricultural Polytechnic Institute (rural vocational high school)
IPCC/AFOS Intergovernmental Panel on Climate Change/Agriculture, Forestry, and Other Systems
IPM integrated pest management
ISACA Advanced Institute of Agricultural Sciences of Ciego de Ávila (now the University of Ciego de Ávila)
ISCAH Advanced Institute for Agricultural Sciences of Havana (now UNAH)
ISCOD Trade Union Institute for Cooperation in Development (Spain)
JNE National Economic Council
LABIOFAM Biological Pharmaceuticals Laboratory
LAPROSAV Provincial Plant Protection Laboratory
LER land equivalent ratio
MAELA Latin American Agroecological Movement
MES Ministry of Higher Education
MINAG Ministry of Agriculture
MINAL Ministry of Food
MINAZ Ministry of Sugar
MINED Ministry of Education
MININT Ministry of the Interior
MINSAP Ministry of Public Health
MINVEC Ministry for Foreign Investment and Economic Collaboration
MVA mycorrhizal fungi
NCOS National Center for Development (Belgium)
NGO nongovernmental organization
OLADE Latin American Energy Organization
ONE National Statistics Office
ONN National Standards Office
OREALC Regional Office for Latin America and the Caribbean, UNESCO
ORLAC Regional Office for Latin America and the Caribbean, FAO
PAN National Food Program

PAN Pesticide Action Network
PCC Cuban Communist Party
RIAD International Network for Agriculture and Democracy
SANE Sustainable Agriculture Networking and Extension program of UNDP
SEDAGRI Agricultural Publications and Design Service
SERFE Fertilizers and Amendments Extension Service
SERVAR Fertilizers and Varieties Service
TCP technical cooperation project
UBPC Basic Units of Cooperative Production
UCC Uruguayan Cooperative Center
UCLV Central University of Las Villas
UNAH Agrarian University of Havana
UNDP United Nations Development Program
UNESCO United Nations Organization for Education, Science, and Culture
UNIDO United Nations Industrial Development Organization

About the Authors

Luis Alemán Manzfarroll. Hydraulic engineer; Ph.D. in Agricultural Sciences (Krasnodar, Russia); Adjunct Professor, University of Camagüey; Researcher and Director, Rice Research Institute, Ministry of Agriculture (MINAG).

Miguel A. Altieri. Agroecologist; Ph.D. in Entomology (USA); Associate Professor, Environmental Science, Policy and Management Department, University of California, Berkeley, California.

Mavis D. Álvarez Licea. Agronomist; Director of International Cooperation, National Association of Small Farmers (ANAP).

Martin Bourque. Ecologist; M.A. in Latin American Studies (University of California at Berkeley); former staff member, Institute for Food and Development Policy (Food First), USA; Executive Director, The Ecology Center, Berkeley, California.

Antonio Casanova Morales. Agronomist; Ph.D. in Agricultural Sciences (Horticultural Institute, Maritza, Bulgaria); Adjunct Professor, Central University of Las Villas; Researcher, Liliana Dimitrova Horticultural Research Institute (IIHLD), Ministry of Agriculture (MINAG).

Nelso Companioni Concepción. Agronomist; Ph.D. in Agricultural Sciences (Russia); Researcher and Deputy Director, National Institute for Fundamental Research on Tropical Agricultural (INIFAT), Ministry of Agriculture (MINAG).

Ricardo Delgado Díaz. Agronomist; Ph.D. in Agricultural Sciences (Cuba); President, Cuban Association of Agricultural and Forestry Technicians (ACTAF).

Dayma Echevarría León. Sociologist; M.A. (University of Havana); Researcher, Rural Studies Team (EER), Department of Sociology, University of Havana.

José Manuel Febles González. Agronomist; Ph.D. in Agricultural Sciences (Cuba); Professor, Center for the Study of Sustainable Agriculture (CEAS), Agrarian University of Havana; Director of International Relations, Agrarian University of Havana (UNAH).

Fernando Funes Aguilar. Agonomist; Diploma in Tropical Agronomy (University of Queensland, Australia); Ph.D. in Agricultural Sciences (Cuba); Adjunct Professor, Agrarian University of Havana; Secretary of the Organic Farming Group (GAO) of the Cuban Association of Agricultural and Forestry Technicians (ACTAF); Deputy Director for Research, Pasture and Forage Research Institute (IIPF), Ministry of Agriculture (MINAG).

Fernando Funes Monzote. Agronomist; M. Sc. (University of La Rábida, Spain); Diploma in Dairy Production and Rural Development (University of Wageningen, Holland); Researcher, Pasture and Forage Research Institute (IIPF), Ministry of Agriculture (MINAG).

Mercedes García Negrín. Chemist; Specialist in Medicinal Plants; Laboratory Director, Pasture and Forage Research Institute (IIPF), Ministry of Agriculture (MINAG).

Luis García Pérez. Agronomist; Ph.D. in Agricultural Sciences (Cuba); Professor and Director, Center for the Study of Sustainable Agriculture (CEAS), Agrarian University of Havana (UNAH).

Margarita García Ramos. Agronomist; Ph.D. in Agricultural Sciences (Cuba), Researcher and Team Leader, Nutritional Alternatives Group, National Institute of Agricultural Sciences (INCA), Ministry of Higher Education (MES).

Adrián Hernández Chávez. Agronomist; M.Sc. in Agroecology and Sustainable Agriculture (Cuba); Researcher and Division Chief, Crop Technology, Liliana Dimitrova Horticultural Research Institute (IIHLD), Ministry of Agriculture (MINAG).

Richard Levins. Ecologist; Ph.D. in Ecology (USA); John Rock Professor of Population Science, Harvard University School of Public Health, Boston, Massachusetts; Adjunct Researcher, Institute of Ecology and Systematics (IES), Havana, Cuba.

Lucy Martín Posada. Sociologist; Associate Researcher, Center for Psychological and Sociological Research (CIPS).

Rafael Martínez Viera. Biologist; Ph.D. in Agricultural Sciences (University of Leipzig, Germany); Professor and Researcher, Agrarian University of Havana (UNAH); Division Chief, Microbiology and Biochemistry, National Institute for Fundamental Research on Tropical Agriculture (INIFAT), Ministry of Agriculture (MINAG).

Marta Monzote Fernández. Biologist; Ph.D. in Agricultural Sciences (Cuba), Researcher, Pasture and Forage Research Institute (IIPF), Ministry of Agriculture (MINAG).

Rafael Morín Pérez. Agronomist; Organic Farming Certifier (AIAB, Italy and ECO-CERT, Europe); Certification Specialist, Certification Department, National Standards Office (ONN).

Eulogio Muñoz Borges. Livestock Specialist and Veterinary Doctor; Adjunct Professor, Agrarian University of Havana (UNAH), Researcher, Animal Science Institute (ICA), Ministry of Higher Education (MES).

Catherine Murphy. Sociologist; M.Sc. in Sociology (FLACSO, Cuba), former staff member, Institute for Food and Development Policy (Food First), USA.

Marcos Nieto Lara. Agronomist and Economist; Head, Negotiating Team, Ministry of Agriculture (MINAG).

Armando Nova González. Economist; Researcher and Professor, Agrarian University of Havana (UNAH), Researcher and Professor, Center for the Study of the Cuban Economy, University of Havana.

Yanet Ojeda Hernández (deceased). Agronomist; Researcher, National Institute for Fundamental Research on Tropical Agriculture (INIFAT), Ministry of Agriculture (MINAG).

Egidio Páez Medina. Agronomist and Plant Protection Specialist; President, Havana City Chapter, Cuban Association of Agricultural and Forestry Technicians (ACTAF).

Nilda Pérez Consuegra. Agronomist and Plant Protection Specialist; M.Sc. in Agroecology and Sustainable Agriculture (Cuba); Assistant Professor, Center for the Study of Sustainable Agriculture (CEAS), Agrarian University of Havana (UNAH).

Niurka Pérez Rojas. Ph.D. in Law, M.Sc. in Sociology (FLACSO-Chile), Ph.D. in History and Sociology (Academy of Sciences, Russia); Team Leader, Rural Studies Team (EER), Sociology Department, University of Havana.

Félix Ponce Ceballos. Mechanical Engineer; Ph.D. in Technical Sciences (Russev, Bulgaria); Professor and Assistant Dean of the College of Agricultural Mechanization, Agrarian University of Havana (UNAH).

Pedro Luis Quintero León. Agronomist; M.Sc. in Agroecology and Sustainable Agriculture (Cuba); Farm Specialist, Self-Provisioning Program, Civilian Aeronautics Institute of Cuba (IACC).

Arcadio Ríos Hernández. Mechanical Engineer; Ph.D. in Technical Sciences (Bulgaria); Adjunct Professor, Agrarian University of Havana; Researcher, Agricultural Mechanization Research Institute (IIMA), Ministry of Agriculture (MINAG).

Peter Rosset. Agroecologist; M.Sc. in Entomology (University of London, UK); Ph.D. in Agricultural Ecology (University of Michigan, USA); Co-Director, Institute for Food and Development Policy (Food First), USA.

Salvador Sánchez Sánchez. Agronomist; Researcher, Rice Research Institute (IIA), Ministry of Agriculture (MINAG).

Miguel Socorro Quesada. Agronomist; Ph.D. in Agricultural Sciences (Kuban, Russia); Adjunct Professor, University of Camagüey; Reseacher and Head of "Popular Rice" Program, Rice Research Institute (IIA), Ministry of Agriculture (MINAG).

Rafael Suárez Rivacoba. Chemist; Researcher; Representative in Brazil of the Ministry of Sugar (MINAZ).

Eolia Treto Hernández. Agronomist; Ph.D. in Agricultural Sciences (Cuba); Adjunct Professor, Agrarian University of Havana (UNAH); Researcher, National Institute of Agricultural Sciences (INCA), Ministry of Higher Education (MES).

Luis L. Vázquez Moreno. Agronomist; Ph.D. in Agricultural Sciences (Cuba); Deputy Director for Science, National Plant Protection Institute (INISAV), Ministry of Agriculture (MINAG).

Directory of Institutions and Authors

Agrarian University of Havana
(Universidad Agraria de La Habana–UNAH)
Autopista Nacional y Carretera de Tapaste, San José de las Lajas, La Habana, Cuba, or
Apartado 18-19, San José de las Lajas, La Habana, Cuba
TEL: +53-64-63013
FAX: +53-7-240942

Agricultural Mechanization Research Institute
(Instituto de Investigaciones de Mecanización Agropecuaria–IIMA)
Carretera de Fontanar km 2½, Abel Santamaría, Boyeros, La Habana, Cuba
TEL: +53-7-451731/453286
FAX: +53-7-335875
EMAIL: iima@ip.etecsa.cu

Animal Science Institute
(Instituto de Ciencia Animal–ICA)
Carretera Central km 47½, San José de las Lajas, or
Apartado 24, San José de las Lajas, La Habana, Cuba
TEL: +53-62-99180/99410
FAX: +53-7-335382
EMAIL: ica@ceniai.inf.cu

Center for Psychological and Sociological Research
(Centro de Investigaciones Psicológicas y Sociológicas–CIPS)
Calle B, # 352, entre 15 y 17, El Vedado, Plaza, Ciudad de La Habana, Cuba
TEL: +53-7-35366/306674
FAX: +53-7-334327
EMAIL: cips@ceniai.inf.cu

Center for the Study of Sustainable Agriculture
(Centro de Estudios de Agricultura Sostenible–CEAS, UNAH)
Apartado 18–19, Autopista Nacional y Carretera de Tapaste, San José de las Lajas, La Habana, Cuba
TEL: +53-64-63013
FAX: +53-7-240942
EMAIL: lantgarcia@hotmail.com
nilda_cu@yahoo.com

Center for the Study of the Cuban Economy
(Centro de Estudios de la Economía Cubana, Universidad de la Habana)
Avenida 41, #707, Esq. a 9na, Miramar, Playa, Ciudad de La Habana, Cuba
TEL: +53-7-221391/290563/240987
FAX: +53-7-330987
EMAIL: ceec@comuh.uh.cu

Civilian Aeronautics Institute of Cuba
(Instituto de Aeronáutica Civil de Cuba–IACC)
Calle 23, #64, La Rampa, El Vedado, Plaza, Ciudad de La Habana, Cuba
TEL: +53-7-551136
FAX: +53-7-334553

College of Agricultural Mechanization
(Facultad de Mecanización Agropecuaria, Universidad Agraria de La Habana)
Apartado 18–19, Autopista Nacional y Carretera de Tapaste, San José de las Lajas, La Habana, Cuba
TEL: +53-64-63013
FAX: +53-7-240942

Cuban Association of Agricultural and Forestry Technicians
(Asociación Cubana de Técnicos Agrícolas y Forestales–ACTAF)
Edificio MINAG Piso 9, Avenida de Independencia y Conill, Plaza, Ciudad de La Habana, Cuba
TEL: +53-7-845387/845266
FAX: +53-7-845387
EMAIL: actaf@minag.gov.cu

Ecology Center
2530 San Pablo Avenue, Berkeley, CA, 94702 USA
TEL: +1-510-548-2220
FAX: +1-510-548-2240
EMAIL: martinazo@ecologycenter.org

Environmental Science, Policy, and Management Department, University of California
201 Wellman Hall, University of California, Berkeley, CA 94720-3112 USA
TEL: +1-510-642-9802
FAX: +1-510-642-3327
EMAIL: agroeco3@nature.berkeley.edu

Harvard School of Public Health
Harvard University, 665 Huntington Avenue, Boston, MA 02115 USA
TEL: +1-617-432-1232
FAX: +1-617-566-0365
EMAIL: humaneco@hsph.harvard.edu

Havana City Chapter, ACTAF
(Filial ACTAF Ciudad de La Habana)
Finca "El Paraíso", Calle 3ra., entre Porvenir y Línea del Ferrocarril, Víbora Park, Arroyo Naranjo, Ciudad de La Habana, Cuba
TEL: +53-7-444940

Institute for Food and Development Policy (Food First)
398 60th Street, Oakland, California, 94618, USA
TEL: +1-510-654-4400
FAX: +1-510-654-4551
EMAIL: foodfirst@foodfirst.org
WEB SITE: www.foodfirst.org

Liliana Dimitrova Horticultural Research Institute
(Instituto de Investigaciones Hortícolas "Liliana Dimitrova"–IIHLD)
Carretera Bejucal–Quivicán km 33½, La Habana, Cuba
TEL: +53-66-82600, +53-67-57755
FAX: +53-66-82601
EMAIL: liliana@colombus.cu

Ministry of Sugar
(Ministerio del Azúcar–MINAZ)
Calle 23, #171, entre N y O, El Vedado, Plaza, Ciudad de La Habana, Cuba
TEL: +53-7-553214

National Association of Small Farmers
(Asociación Nacional de Agricultores Pequeños–ANAP)
Calle I, # 206, entre Línea y 13, El Vedado, Ciudad de La Habana, Cuba
TEL: +53-7-328586/324541
FAX: +53-7-328586/333044/334244
EMAIL: amigo@anap.org.cu

National Institute for Fundamental Research on Tropical Agriculture
(Instituto de Investigaciones Fundamentales en Agricultura Tropical–INIFAT)
Calle 1, esquina a 2, Santiago de las Vegas, Boyeros, Ciudad de La Habana, Cuba
TEL: +53-7-683-4039/2323, +53-7-579010
FAX: +53-7-579014
EMAIL: inifat@ceniai.inf.cu

National Institute of Agricultural Sciences
(Instituto Nacional de Ciencias Agrícolas–INCA)
Gaveta Postal 1, Código Postal 32700, San José de las Lajas, La Habana, Cuba, or Autopista Nacional y Carretera de Tapaste, San José de las Lajas, La Habana, Cuba
TEL: +53-64-63867
FAX: +53-64-63807
EMAIL: dir@inca.edu.cu

National Plant Protection Institute
(Instituto de Investigaciones de Sanidad Vegetal–INISAV)
Calle 110, #514, entre 5ª b y 5ª F, Miramar, Playa, Ciudad de La Habana, Cuba
TEL: +53-7-233907/222510/296189/226788
FAX: +53-7-280535/283703
EMAIL: inisav@ceniai.inf.cu

National Standards Office
(Oficina Nacional de Normalización)
Calle E, #261, entre 11 y 13, El Vedado, Plaza, Ciudad de La Habana, Cuba
TEL: +53-7-300825/300835
FAX: +53-7-300588
EMAIL: ncnorma@ceniai.inf.cu

Negotiating Team, Ministry of Agriculture
(Grupo Negociador, MINAG)
Edificio MINAGRI Piso 6, Avenida de Independencia y Conill, Plaza, Ciudad de La Habana, Cuba
TEL: +53-7-845377/845305/845500
FAX: +53-7-845243
EMAIL: nieto@minag.gov.cu

Pasture and Forage Research Institute
(Instituto de Investigaciones de Pastos y Forrajes–IIPF)
Carretera 43 km 1½, Cangrejeras, Bauta, La Habana, Cuba
TEL: +53-7-299855, +53-680-3559
TEL/FAX: +53-7-299855
EMAIL: iipf@ceniai.inf.cu

Rice Research Institute
(Instituto de Investigaciones del Arroz)
Autopista Novia del Mediodía, km 16½, Bauta, La Habana, Cuba
TEL: +53-680-3550/3260

Rural Studies Team, University of Havana
(Equipo de Estudios Rurales–EER, Departamento de Sociología, Universidad de la Habana)
Facultad de Filosofía e Historia, Universidad de La Habana, San Lázaro, esquina a L, El Vedado, Plaza, Ciudad de La Habana, Cuba
TEL: +53-7-703355
FAX: +53-7-335774
EMAIL: niurka@cubarte.cult.cu

Index

D

E

J

FOOD FIRST BOOKS OF RELATED INTEREST

Basta! Land and the Zapatista Rebellion in Chiapas
Revised edition
George A. Collier with Elizabeth Lowery Quaratiello
Foreword by Peter Rosset

The classic on the Zapatista's in a new revised edition, including a preface by Roldolfo Stavenhagen, a new epilogue about the present challenges to the indigenous movement in Chiapas, and an updated bibliography.
Paperback, $14.95

The Future in the Balance: Essays on Globalization and Resistance
Walden Bello
Edited with a preface by Anuradha Mittal

A new collection of essays by Third World activist and scholar Walden Bello on the myths of development as prescribed by the World Trade Organization and other institutions, and the possibility of another world based on fairness and justice.
Paperback, $13.95

Views from the South: The Effects of Globalization and the WTO on Third World Countries
Foreword by Jerry Mander
Afterword by Anuradha Mittal
Edited by Sarah Anderson

This rare collection of essays by Third World activists and scholars describes in pointed detail the effects of the WTO and other Bretton Woods institutions.
Paperback, $12.95

America Needs Human Rights
Edited by Anuradha Mittal and Peter Rosset

This new anthology includes writings on understanding human rights, poverty in America, and welfare reform and human rights.
Paperback, $13.95

The Paradox of Plenty: Hunger in a Bountiful World
Edited by Douglas H. Boucher

Excerpts from Food First's best writings on world hunger and what we can do to change it.
Paperback, $18.95

A Siamese Tragedy: Development and Disintegration in Modern Thailand
Walden Bello, Shea Cunningham, and Li Kheng Poh

Critiques the failing economic system that has propelled the Thai people down an unsustainable path.
Paperback, $19.95

Dark Victory: The US and Global Poverty
Walden Bello, with Shea Cunningham and Bill Rau
Second edition, with a new epilogue by the author

Offers an understanding of why poverty has deepened in many countries, and analyzes the impact of US economic policies.
Paperback, $14.95

Dragons in Distress: Asia's Miracle Economies in Crisis
Walden Bello and Stephanie Rosenfeld

After three decades of rapid growth, the economies of South Korea, Taiwan, and Singapore are in crisis. The authors offer policy recommendations to break these countries from their unhealthy dependence on Japan and the US.
Paperback, $12.95

Education for Action: Graduate Studies with a Focus on Social Change
Fourth edition
Edited by Joan Powell

A newly updated authoritative and easy-to-use guidebook that provides information on progressive programs in a wide variety of fields.
Paperback, $12.95

Alternatives to the Peace Corps: Third World and US Volunteer Opportunities
Ninth edition
Edited by Joan Powell

Over one hundred listings of organizations in the United States and the Third World provide the prospective volunteer an array of choices to make their commitment count.
Paperback, $9.95

Write or call our distributor to place book orders. All orders must be prepaid. Please add $4.50 for the first book and $1.50 for each additional book for shipping and handling.

LPC Group
22 Broad Street, Suite 22
Milford, CT 06460
www.coolbooks.com (800) 343-4499

ABOUT FOOD FIRST

(Institute for Food and Development Policy)

Food First, also known as the Institute for Food and Development Policy, is a nonprofit research and education-for-action center dedicated to investigating and exposing the root causes of hunger in a world of plenty. It was founded in 1975 by Frances Moore Lappé, author of the bestseller *Diet for a Small Planet,* and food policy analyst Dr. Joseph Collins. Food First research has revealed that hunger is created by concentrated economic and political power, not by scarcity. Resources and decision-making are in the hands of a wealthy few, depriving the majority of land, jobs, and therefore food.

Hailed by *The New York Times* as "one of the most established food think tanks in the country," Food First has grown to profoundly shape the debate about hunger and development.

But Food First is more than a think tank. Through books, reports, videos, media appearances, and speaking engagements, Food First experts not only reveal the often hidden roots of hunger, they show how individuals can get involved in bringing an end to the problem. Food First inspires action by bringing to light the courageous efforts of people around the world who are creating farming and food systems that truly meet people's needs.

HOW TO BECOME A MEMBER OR INTERN OF FOOD FIRST

BECOME A MEMBER OF FOOD FIRST

Private contributions and membership gifts form the financial base of Food First/Institute for Food and Development Policy. The success of the Institute's programs depends not only on its dedicated volunteers and staff, but on financial activists as well. Each member strengthens Food First's efforts to change a hungry world. We invite you to join Food First. As a member you will receive a twenty percent discount on all Food First books. You will also receive our quarterly publication, *Food First News and Views*, and timely *Backgrounders* that provide information and suggestions for action on current food and hunger crises in the United States and around the world. If you want so subscribe to our internet newsletter, *Food Rights Watch*, send us an e-mail at foodfirst@foodfirst.org. All contributions are tax-deductible.

BECOME AN INTERN FOR FOOD FIRST

There are opportunities for interns in research, advocacy, campaigning, publishing, computers, media, and publicity at Food First. Our interns come from around the world. They are a vital part of our organization and make our work possible.

To become a member or apply to become an intern, just call, visit our web site, or clip and return the attached coupon to

Food First/Institute for Food and Development Policy
398 60th Street, Oakland, CA 94618, USA
PHONE: (510) 654-4400 FAX: (510) 654-4551
EMAIL: foodfirst@foodfirst.org
WEB SITE: www.foodfirst.org

You are also invited to give a gift membership to others interested in the fight to end hunger.

JOINING FOOD FIRST

☐ I want to join Food First and receive a 20% discount on this and all subsequent orders. Enclosed is my tax-deductible contibution of:

☐ $35 ☐ $50 ☐ $100 ☐ $500 ☐ $1,000 ☐ OTHER

NAME ____________________

ADDRESS ____________________

CITY/STATE/ZIP ____________________

DAYTIME PHONE (____) ____________________

E-MAIL ____________________

ORDERING FOOD FIRST MATERIALS

ITEM DESCRIPTION	QTY	UNIT COST	TOTAL

PAYMENT METHOD:

☐ CHECK

☐ MONEY ORDER

☐ MASTERCARD

☐ VISA

MEMBER DISCOUNT, 20% $ ______

CA RESIDENTS SALES TAX 8.25% $ ______

SUBTOTAL $ ______

POSTAGE 15% UPS: 20% ($2 MIN.) $ ______

MEMBERSHIP(S) $ ______

ADDITIONAL CONTRIBUTION $ ______

TOTAL ENCLOSED $ ______

____

NAME ON CARD

____

CARD NUMBER EXP. DATE

____

SIGNATURE

MAKE CHECK OR MONEY ORDER PAYABLE TO:

Food First, 398 - 60th Street, Oakland, CA 94618

FOR GIFT MEMBERSHIPS & MAILINGS, PLEASE SEE COUPON ON REVERSE SIDE

FOOD FIRST GIFT BOOKS

Please send a Gift Book to (order form on reverse side):

NAME ______________________________

ADDRESS ______________________________

CITY/STATE/ZIP ______________________________

FROM ______________________________

FOOD FIRST PUBLICATIONS CATALOGS

Please send a Publications Catalog to:

NAME ______________________________

ADDRESS ______________________________

CITY/STATE/ZIP ______________________________

NAME ______________________________

ADDRESS ______________________________

CITY/STATE/ZIP ______________________________

NAME ______________________________

ADDRESS ______________________________

CITY/STATE/ZIP ______________________________

FOOD FIRST GIFT MEMBERSHIPS

☐ Enclosed is my tax-deductible contribution of:

☐ $35 ☐ $50 ☐ $100 ☐ $500 ☐ $1,000 ☐ OTHER

Please send a Food First membership to:

NAME ______________________________

ADDRESS ______________________________

CITY/STATE/ZIP ______________________________

FROM ______________________________